RAND NATIONAL DEFENSE RESEARCH INSTITUTE

T0127926

360-Degree Assessments

Are They the Right Tool for the U.S. Military?

Chaitra M. Hardison, Mikhail Zaydman, Tobi Oluwatola,
Anna Rosefsky Saavedra, Thomas Bush, Heather Peterson,
Susan G. Straus

Prepared for the Office of the Secretary of Defense

For more information on this publication, visit www.rand.org/t/RR998

Library of Congress Cataloging-in-Publication Data is available for this publication

ISBN: 9780-8330-8905-2

Published by the RAND Corporation, Santa Monica, Calif.

© Copyright 2015 RAND Corporation

RAND® is a registered trademark.

Support RAND

Make a tax-deductible charitable contribution at

www.rand.org/giving/contribute

www.rand.org

Preface

The National Defense Authorization Act for Fiscal Year 2014 directed the Secretary of Defense to assess "the feasibility of including a 360-degree assessment approach . . . as part of performance evaluation reports" and to report to Congress on the findings of that assessment. The Office of the Under Secretary of Defense for Personnel and Readiness (OUSD/P&R) asked the RAND Corporation to assist with this assessment, and this report provides OUSD/P&R with information relevant to that request. OUSD/P&R asked RAND not only to explore the use of 360-degree assessments (360s) in the military for evaluation purposes, as requested by Congress, but also to provide information on the role of 360s more broadly. Thus, our study explores the pros and cons of using 360s for evaluation and development purposes in the military.

Our research is based on information gleaned from a number of sources: existing research literature and expert guidance on 360 best practices; policy documents and other sources summarizing current performance and promotion practices in the military services, including the use of 360s; and interviews with a sample of stakeholders and subject-matter experts in the Department of Defense. The results of our research suggest that using 360 feedback as part of the military performance evaluation system is not advisable at this time, though the services could benefit from using 360s as a tool for leader development and to gain an aggregate view of leadership across the force.

The research reported here will be of interest to individuals in the services, Congress, and the Department of Defense who are responsible for military personnel policies, or who are considering implementing 360 feedback with military personnel.

This research was sponsored by the Office of the Under Secretary of Defense for Personnel and Readiness and conducted within the Forces and Resources Policy Center of the RAND National Defense Research Institute, a federally funded research and development center sponsored by the Office of the Secretary of Defense, the Joint Staff, the Unified Combatant Commands, the Navy, the Marine Corps, the defense agencies, and the defense Intelligence Community.

For more information on the RAND Forces and Resources Policy Center, see http://www.rand.org/nsrd/ndri/centers/frp.html or contact the director (contact information is provided on the web page).

Contents

Figures and Tables

Summary

360-degree assessments (360s)—in which employees receive feedback from supervisors, subordinates, and peers—have become popular personnel management tools. The most common use of 360s is as a feedback and development tool, and many civilian organizations have embraced 360s as one of several ways to develop their workforce. Likewise, the military has also begun exploring the use of 360s as a way to contribute to personnel development, most commonly for officers in leadership positions.

In spite of their popularity, however, 360s are not a panacea. Even when used for development purposes, the use of 360s can face legitimate criticism. Given that 360s can be time-consuming to complete and their effectiveness can vary, some worry that their use could be a significant waste of resources. Moreover, experts also caution that, when used in a way in which the results will ultimately impact high-stakes career outcomes (such as promotions), 360s can lead to distrust and have the potential to do more harm than good. Although many of these arguments against using 360s are legitimate, some could be mitigated with careful attention to the design of the 360. Others, however, should lead users to think carefully about when it is worthwhile to implement 360s and when it is not.

To better understand these criticisms and other pros and cons of using 360s in a military context, the Office of the Under Secretary of Defense for Personnel and Readiness (OUSD/P&R) reached out to the RAND Corporation for assistance. Part of the impetus for OUSD/P&R's request was the need to respond to section 571 of the National Defense Authorization Act (NDAA) for Fiscal Year 2014, which required the Secretary of Defense to

> submit to the Committees on Armed Services of the Senate and the House of Representatives a report containing the results of an assessment of the feasibility of including a 360-degree assessment approach, modeled after the current Department of the Army Multi-Source Assessment and Feedback (MSAF) Program, as part of performance evaluation reports.

This report provides OUSD/P&R with information relevant to that request. OUSD/P&R asked RAND to explore the use of 360s in the military for evaluation purposes, as instructed in the NDAA, but also asked RAND for assistance in understanding the role of 360 feedback more broadly. As a result, this study explores not only the pros and cons of using 360s for evaluative purposes in the military but also the pros and cons of its use for developmental purposes.

Research Questions and Approach

We developed four overarching questions to help guide the department's understanding of major policy issues and impacts when implementing 360s in the services:

- What is known about 360-degree assessments?
- Are the military services already using 360s? If so, how are they using them?
- Would it be advisable to implement 360s for development or evaluation purposes for all officers[1] in the military? Why or why not?
- What implementation challenges should the services be aware of to ensure success?

The information supporting our analysis is drawn from several sources: (1) existing research on organizational experience with 360s; (2) documentation, policies, and statutes on promotions and performance evaluations in the military services; and (3) structured interviews with personnel knowledgeable about the largest 360s currently in use in the services and with key stakeholders and military personnel groups from across the services, the Joint Staff, and the Office of the Secretary of Defense. Each of these sources provides a unique but relevant contribution to the report.

The Great 360 Debate

The debate about 360s centers on how 360 feedback is used—specifically, whether it should be used for developmental purposes, for evaluation purposes, or both. Research on 360 feedback programs suggest that the vast majority of companies that implement 360s use them for development-related purposes, such as training, annual development goals, and succession planning. Few use 360 results solely for evaluation purposes that affect personnel decisions, such as promotions and salary increases.

To the degree that the military currently uses 360s, the purpose is entirely for individual development. Each of the military services and the Joint Staff uses some version of 360-degree assessments. The Army, with the most experience by far, has spent 14 years developing its Multi-Source Assessment and Feedback program, which now has widespread implementation. In the other services, the use of 360s is more targeted—generally directed at senior leadership or toward high-potential officers as part of the military education system. A number of the programs are in early stages of development and are undergoing pilot tests on small subsets of the population. All of the programs are developmental in nature, and their small size allows for personalized coaching support in the interpretation and utilization of results. To date, none of the services uses 360s as an evaluation tool in the performance appraisal process.

Should the military services consider broader use and implement 360s as part of performance evaluation? Should the use of 360s be mandated, even if only for developmental purposes? Are the military services on track? Our discussions with stakeholders throughout the Department of Defense resulted in a fairly consistent response to these questions.

---

[1]   The Office of the Secretary of Defense suggested that we focus on active-duty officers only, with the acknowledgment that the question of 360 uses could certainly be relevant to other military populations as well (civilians, enlisted, reserve, and guard).

"No" for Evaluation

Based on our research both within and outside the military setting, we advise against incorporating 360s in the officer evaluation system at this time. Although it might be *technically* feasible, we advise against it for several reasons. For one, it could ruin developmental applications, and the Army's experience shows how long it can take to develop trust in such a process. Trying to implement two systems could be confusing to raters and increase survey burden on the force. Because of the complexity of using 360s in the evaluation system, mandating their use could lead to largely useless or meaningless ratings that are difficult to interpret without considerable context. Perhaps the key concern in using 360s beyond developmental purposes is their impact on selection boards and the promotion process. Under current law, if 360s are included in an officer's formal evaluation report, the results of that assessment would be part of the documentation provided to selection boards. That said, the services could consider other alternatives for incorporating a broader range of perspectives, including from peers and subordinates, into the performance evaluation system—though identifying specific alternatives is beyond the scope of this study.

"Yes" for Development, but Proceed with Care

We do advise the use of 360s for development purposes for people at higher grades or in leadership positions. We target leadership positions because 360s require a significant investment in time from service members and from raters, and the benefits of 360 feedback for those in leadership positions are most obvious. The tool could be made available as a service to individuals hoping to improve, along with coaching to help service members evaluate the results and incorporate them into self-improvement goals. 360s could also be used to provide an aggregate view of leadership performance across the force—something that current tools are not necessarily well positioned to provide. Leaders could identify locations with leadership issues and could use 360 results to identify force-wide strengths and weaknesses. The bottom line is that making 360 feedback available for developmental use in the military services is a good idea and is essentially how the tool is being used today. Each of the services uses 360 assessments in ways that support their personnel management goals and are in line with service culture.

Current Implementation Is on the Right Track

Overall, our interviews showed that the spirit of 360 clearly resonates with the services. The services value good leadership behaviors and tools that can help develop good leaders. 360s are one tool that has value in developing leaders. Despite their utility, however, 360s are not a silver bullet. Experts in the field of personnel research agree and caution that overzealous use of 360s without careful attention to content, design, and delivery can be harmful to an organization and not worth the time and money spent on them. Thus, mandating 360 assessments force-wide, even for development purposes, is not necessarily the right answer to solving leadership problems within the services and could waste significant resources. Rather it is more advisable to allow the services to continue on their current paths, expanding the use of 360s in a way that is tailored to individual service needs and goals.

Acknowledgments

The authors would like to express sincere gratitude to all of the interview participants, listed by name in the first chapter of this report, who graciously offered their time and insights to the study. The report also benefited greatly from the assistance of Barbara Bicksler, who worked tirelessly providing expert communication and writing assistance on numerous drafts and sponsor briefings. We also thank Lisa Harrington and Maria Lytell, who provided thoughtful comments on early drafts of this report.

CHAPTER ONE

Introduction

360-degree assessments (360s)—in which employees receive feedback from supervisors, subordinates, and peers—have grown to be popular tools in the private sector. The most common use of the 360 is as a feedback and development tool. Because performance improvement requires feedback about one's strengths and weaknesses, offering insights into how subordinates, peers, and others view an individual could be a useful and powerful tool for employee development. As a result, many civilian organizations have embraced 360s as one of several ways to develop their workforce.

For many of the same reasons, the military has also begun exploring the use of 360s as a way to contribute to service member development, particularly officers in leadership positions. For example, the Army currently uses the Army Multi-Source Assessment and Feedback (MSAF) as a mandatory developmental tool for all of its officers and as an optional developmental tool for its civilian and enlisted personnel. That effort, which started as a small pilot in the early 2000s, has grown over the past 14 years to the large-scale effort in place today. Other military services and the Joint Staff are exploring the use of 360s for similar purposes, although some of their implementations are still largely in their infancy and most are on a much smaller scale.

In spite of their popularity, however, 360s are not a panacea. Even when used for development purposes, they can face legitimate criticism. Because 360s can be time consuming to complete and their effectiveness can vary, some worry that the use of 360s could be a significant waste of resources. Moreover, experts also caution that, when used in a way in which the results will ultimately impact high-stakes career outcomes (such as promotions), they can lead to distrust and have the potential to do more harm than good. Although many of these arguments against using 360s are legitimate, some could be mitigated with careful attention to how the tool is designed. Others, however, should lead users to think carefully about when it is worthwhile to implement it and when it is not.

To better understand these criticisms and other pros and cons of using 360s in a military context, the Office of the Under Secretary of Defense for Personnel and Readiness (OUSD/P&R) reached out to the RAND Corporation for assistance. Part of the impetus for this request was the need to respond to section 571 of the National Defense Authorization Act (NDAA) for Fiscal Year 2014, which required the Secretary of Defense to

> submit to the Committees on Armed Services of the Senate and the House of Representatives a report containing the results of an assessment of the feasibility of including a 360-degree assessment approach, modeled after the current Department of the Army Multi-Source Assessment and Feedback (MSAF) Program, as part of performance evaluation reports.

This report is therefore designed to provide OUSD/P&R with information relevant to that request. OUSD/P&R asked RAND to explore the use of the 360s in the military for *evaluation purposes* specifically because the NDAA requested it, and it asked RAND to provide greater detail on the existing MSAF program in response to its specific mention in the NDAA.

In addition, OUSD/P&R asked RAND for assistance in understanding the role of 360 feedback more broadly. As a result, we expanded the goal of the present research effort to also address whether 360s would be a useful tool for the military for *developmental purposes*. Our study was therefore designed to explore not only the pros and cons of using 360s for evaluative purposes in the military but also the pros and cons of its use for developmental purposes.

Research Questions

There are many questions germane to the feasibility and potential success of using 360s for military personnel. However, answers necessarily depend on the stated purpose and the focal population for the 360. As a result, answering every question relevant to implementing 360s for all possible uses in the military was far beyond the scope of this study. Instead, we focused our questions on broad topics intended to help guide the Office of the Secretary of Defense's (OSD) understanding of the major policy issues and impacts when implementing 360s in the services.

Because the NDAA language seemed to imply a force-wide implementation of 360s, we consulted with OUSD/P&R to determine how best to scope the population of interest for our research questions. Based on that guidance, we focused our efforts on active-duty officers only, acknowledging that the use of 360s could be relevant to other military populations as well (civilians, enlisted, reserve, and guard).

In consultation with our sponsor, we identified the following broad questions as the primary focus of our efforts:

- What is known about 360-degree assessments?
- Are the military services already using 360s? If so, how are they using them?
- Would it be advisable to implement 360s for developmental or evaluation purposes for all officers in the military? Why or why not?
- What implementation challenges should the services be aware of to ensure success?

Method

Our analysis pulled information from several sources to address these questions. We conducted structured interviews with developers of 360s currently in use in the services and with key stakeholders and military personnel groups. We drew on existing research and expert advice on organizational experience with 360-degree assessment, examining such issues as how experts recommend using this tool, reactions to 360 feedback, the accuracy of 360 feedback, and effects of 360 feedback on subsequent performance. Finally, we reviewed documentation, policies, and statutes on promotions and performance evaluations in the military services to learn about the military's current use of 360s and how 360s would complement or duplicate existing processes. Each of these sources provides a unique but relevant contribution to this report.

In the sections that follow, we pull from these sources in responding to the questions posed, though not all sources are relevant for addressing every topic area.

Stakeholders and Subject-Matter Expert Interviews

The purpose of our interviews was to collect input from the policymakers involved in personnel evaluation or administration of existing 360s. Toward that end, the project sponsor identified 13 individuals for us to interview from across the services, the Joint Staff, and OSD. Though limited in number, the individuals interviewed are senior policy stakeholders and therefore contribute expert opinion to the evaluation, assessment, promotion, or development process. In addition to interviewing ten of the original 13 individuals, we conducted eight interviews with other subject-matter experts recommended by the original 13, who could help shed light on specific implementation challenges and benefits.

In total, we conducted interviews with 19 distinct groups during the project time frame, as shown in Table 1.1. Each interview included the primary contact and often one or more colleagues invited by the primary contact.

Interviews were semistructured and covered the following questions:

- What types of 360-degree assessment activities is your service or organization currently using? Are they for development or evaluation purposes?
- If the military were to adopt 360-degree assessment for evaluation purposes, what do you foresee as the possible positive and negative consequences?
- How about for development purposes?
- How feasible would it be to implement this for all officers? Are there logistical concerns (stakeholder issues, technology issues, policy and legal impacts, financial costs or other resource issues, etc.)?
- How advisable would it be for the military to implement 360-degree assessment for development and/or evaluation?
- What else should we consider?

However, we considered the discussions exploratory and therefore probed with additional related questions as relevant.

We analyzed interview comments in two ways. First, we identified a series of comments raised frequently during the discussions and provided counts of the frequency to highlight the most prominent concerns of the group.[1] However, because each participant brought unique perspectives, experiences, and insights to the study, there was also value in conveying a comprehensive picture of all of the ideas and thoughts that were raised, regardless of the frequency. These less frequent comments were particularly insightful in that they identified issues that appear to be legitimate concerns that others had perhaps not previously considered. We therefore summarize their comments and offer paraphrased quotes from our notes to illustrate a number of these points throughout the report.

Although we designed our interviews to focus on policymakers with influence over or responsibility for 360 policies in the services, we acknowledge that they are not the only service members and military leaders who might have opinions about 360s. For example, the

---

[1] Three RAND researchers coded the interview notes and iteratively modified the major categories and subcategories to reach agreement on a final set of coding categories.

Table 1.1
Interview Participants

Group	Name and Affiliation
OSD	RADM Margaret D. (Peg) Klein, Senior Advisor to the Secretary of Defense Lt Col Kevin Basik (U.S. Air Force representative) COL Robert Taradash (U.S. Army representative) LtCol. Darrell Platz (U.S. Marine Corps representative)
	Steven Strong and James Schwenk, Office of the General Counsel
Joint Staff	Leslie A. Purser, Joint Education Advisor, J7 Division Lt Col Paul Amrhein
	BG Margaret W. Burcham, Director for Manpower and Personnel, J1
	LtGen Thomas Waldhauser, Director for Joint Force Development, J7
	COL Richard M. (Mike) Cabrey, Chief of Staff to the Director, Joint Force Development, J7
	Jerome M. Lynes, Deputy Director for Joint Education and Doctrine, Joint Chiefs of Staff
Army	COL Christopher D. Croft, Director, Center for Army Leadership Dr. Jon Fallesen Anthony J. Gasbarre Dr. Melissa Wolfe Clark Delavan LTC Chris Taylor
	LTC Johnny Oliver (representing the Office of MG Thomas Seamands, Director of Military Personnel Management, Army G-1) COL Robert Bennett COL Michael E. Masley Albert Eggerton
	COL Joseph R. Callaway, General Officer Management Office
	Anthony J. Stamilio, Deputy Assistant Secretary of the Army
Air Force	Brig Gen Brian T. Kelly, Director, Force Management Policy
	Jeffrey Mayo, Deputy Assistant Secretary of the Air Force for Force Management Integration
	Russell J. Frasz, Director, Force Development
	Col Christopher E. Craige, General Officer Management Office
Marine Corps	Col. Scott Erdelatz, Director, Lejeune Leadership Institute
Navy	RADM Frederick Roegge, Director, Military Personnel Plans and Policy (N13) CAPT Jeffrey Krusling
	CDR Nicole DeRamus-Suazo, Ph.D., Deputy Director for Navy Flag Officer Development
N/A	LTG Walt Ulmer (Ret.), former Director, Center for Creative Leadership

broader force may strongly desire opportunities to provide upward feedback and may believe that doing so would be valuable. Junior officers in particular might want such an avenue to voice anonymous opinions. Others in the policy community also have views on 360s. In their report reviewing the 2010 Quadrennial Defense Review, Hadley and Perry (2010) recommended that Congress mandate 360s for all officers in the services. That said, the interviewees who were invited to participate provide important perspectives from policymakers currently in a position to shape future efforts to implement 360s on a larger scale.

Research and Expert Advice on 360s

The literature on 360s is expansive. Hundreds of journal articles and dozens of books provide expert guidance on 360 best practices, most published in the 1990s and 2000s. They debate the pros and cons of using 360s, discuss the impacts of 360s on workplace performance, and address issues of when and how to design 360s. Because a number of excellent comprehensive reviews of the existing research can be found elsewhere (see, for example, *The Handbook of Multisource Feedback* by Bracken, Timmreck, and Church, 2001), and because the literature itself is vast, our intention was not to provide yet another comprehensive review. Instead, we sought to use the literature to accomplish two goals. First we aimed to provide a quick summary of 360s as general background information for our readers. Second, and more importantly, where relevant, we pulled in existing research and expert advice to help support or refute some of the issues raised in our interviews with the key stakeholders and military representatives.

Our review therefore focused solely on findings that could inform the decision to adopt a 360-feedback program and factors to consider in program design and implementation. We intentionally omitted some heavily researched areas of 360s that may be less relevant to policymakers. For example, there is an abundant body of literature on individual difference characteristics associated with 360 ratings.[2] While it might be important for coaches and mentors to be aware of how these characteristics can influence ratings in order to help put the ratings in context or to help maximize the usefulness for some individuals, many of the findings point to biological and individual characteristics that are stable and that an organization cannot affect on a large scale. Instead, we focused on factors that organizations are in a position to influence and that are relevant for establishing policy, such as how 360 programs are designed and how the information is used.

Organization of This Report

The remainder of this report responds to the aforementioned research questions. In Chapter Two, we provide a brief overview of 360s as background for our readers. Because the NDAA language proposes using 360s for evaluation purposes, Chapter Three summarizes the existing performance evaluation and promotion process used by the military services at the time of this study. The next chapter, Chapter Four, offers an overview of the military's current use of 360s (in use or in development at the time of the study), with special attention to the Army's MSAF. Chapters Five and Six summarize the key concerns raised by stakeholders regarding implementation of 360s and discuss existing research that addresses those concerns. The final chapter discusses implications of the study's findings and provides recommendations for how 360s should be used in the military services.

---

[2]   Fleenor et al. (2010) contains a comprehensive summary of research on congruence of self and other rating and review findings from numerous studies on effects of biographical characteristics, personality traits, and other dispositional characteristics, as well as job-relevant experience. Ostroff, Atwater, and Feinberg (2004) also reviewed the influence on ratings of rater–ratee similarities or differences in characteristics, such as age, gender, and race. Others have studied individual characteristics, such as personality and dispositional traits, on acceptance of 360 feedback (Smither, London, and Richmond, 2005), goal commitment following 360 feedback (Bono and Colbert, 2005), and performance improvement following 360 feedback (e.g., Heslin and Latham, 2004; Smither, London, and Richmond, 2005).

CHAPTER TWO

What Are 360-Degree Assessments, and How Are They Different from Other Assessments?

Put simply, 360s are a tool that can be used to measure an employee's performance. Like any other tool used for this purpose (such as a multiple-choice test, simulation activity, or supervisory ratings), the method itself is not inherently good or bad. Instead, the desired goal for the tool, what it is intended to measure, how well it is designed, and how it is used can all impact its usefulness.

Supervisor ratings are the most common way in which employee performance is measured. Although a supervisor may solicit input from others in making his or her ratings, in most cases, only the direct supervisor determines a subordinate's rating. Supervisor ratings are typically collected as part of an employee's annual performance evaluation and are used to determine salary actions or promotions, to officially document inadequate performance, and to provide employees with feedback on how their supervisors think they are performing. In many organizations, the annual performance review includes a developmental component in which supervisors may use the formal evaluation as an opportunity to provide the employee with developmental feedback and coaching.

However, in many jobs, a direct supervisor is not the only person who is in a position to comment on a person's performance. Moreover, in some jobs, the direct supervisor may not even be the best situated to provide an accurate assessment of certain aspects of performance. For example, a supervisor may not be in a position to see how someone interacts with the people below him or her on a day-to-day basis. As a result, many experts in personnel evaluation have noted that supervisor ratings can sometimes provide incomplete pictures of employee performance (Hedge, Borman, and Birkeland, 2001; Lepsinger and Lucia, 2001).[1] The 360—which includes self-ratings and feedback from peers, subordinates, and others (such as customers) in addition to the supervisor—is one tool designed to help fill in some of the gaps in supervisor ratings.

Also known as *multisource* or *multirater assessments*, it is this 360-degree view of an employee's performance (illustrated in Figure 2.1) that has led to its most widely recognized name: *360-degree assessment*. In keeping with its most well-recognized name, in this report, we refer to it as *360-degree assessment* or, more simply, *360*. There are many variants on the concept of 360-degree feedback that differ from that illustrated in the figure. Some assessments include only supervisors and subordinates; others include only views from supervisors and customers. Although these variants may be distinguished from 360s by some other authors (see, for

---

[1]   Many studies have also examined possible sources of unreliability in supervisory ratings. For examples, see Landy and Farr, 1980; Murphy, 2008; and Judge and Ferris, 1993.

Figure 2.1
Illustration of 360-Degree Assessment

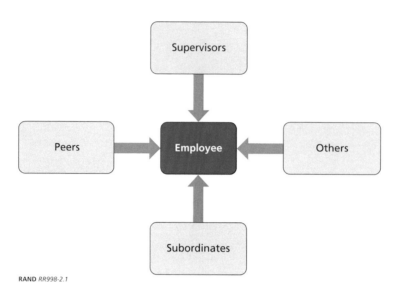

RAND *RR998-2.1*

example, Foster and Law, 2006), *360s* in this report (except where we specify otherwise) refers to assessments that include any source in addition to or instead of supervisors.

360s most commonly measure interpersonal competencies or "soft skills," such as leadership, teamwork, or customer service, that are valued aspects of performance in business settings. Jobs for which a central aspect of performance involves managing project teams, supervising or leading others, or other interactions with people internal or external to the organization are usually well suited to 360 measurements. For those employees, performance on the job may be synonymous with leadership performance and teamwork. However, 360s are not well suited for measuring all aspects of performance in all jobs, such as employees who work in isolation (assembly line factory workers, for example).

Arguments exist for and against using 360s in different contexts. For example, proponents of the 360 argue that it provides a more accurate picture of individual performance than traditional feedback systems because it includes multiple perspectives of employees' performance (e.g., Borman, 1997; Tornow, 1993). This is particularly useful for aspects of performance that, like interpersonal skills, cannot be assessed through purely objective measures, such as the number of items sold or produced. In addition, ratings from multiple raters can be more reliable than ratings from a single rater and therefore may be more valid. 360s can also reduce rater biases and random error (Mount et al., 1998). Proponents of 360 feedback also speak to its value in increasing communication among employees and fostering an increased sense of participation and involvement in the organization that is consistent with the use of teams and other participative management strategies. 360s also can promote a sense of fairness or organizational justice (e.g., Edwards, Ewen, and Vendantam, 2001).

However, there also are costs to 360s. Such programs are resource-intensive to design, implement, and maintain. 360 reviews are time-consuming for participants, who may be reluctant to give feedback or be forthright in their feedback, particularly if they are concerned that their comments can be identified. Additionally, critics note that 360s have, in many cases, been shown to be ineffective at improving behavior and, in some circumstances, can even harm performance.

How Organizations Use 360-Feedback Programs

A recent survey of 360-feedback program managers from 211 organizations[2] in the United States and Canada showed that the use of 360 feedback varies, but the vast majority of those surveyed use it for development-related purposes (Figure 2.2). About one-third use 360s for self-directed development, another third use it for development planning (such activities as identifying training and development needs, annual development goals, and succession planning), and another third use it to make both development and personnel decisions. Few of the companies surveyed use 360 results solely for "personnel decisions such as pay, promotion, and hiring" (3D Group, 2013).

Many researchers have described the pros and cons of using 360s for evaluation purposes versus developmental purposes, with most arguing that they are best used for development only (see London, 2001, for example). The notion that 360s provide better information because peers and subordinates are often in a better position than supervisors to evaluate some behaviors is common to both sides of the debate. 360s can enhance self-development and provide a sound basis for decisions. Arguments in favor of using 360 feedback for evaluative decisions contend that organizations should take advantage of all possible benefits from their investment in 360 programs. On the developmental side, proponents argue that 360s enhance self-awareness, which in turn contributes to behavior change, and that individuals value others' opinions and feel responsible to respond to them. When ratings are anonymous, which is typically the practice in 360s, raters are more likely to be forthright in their feedback.

Experts have offered numerous reasons explaining why using 360 feedback for evaluation decisions is counterproductive. For example, London (2001) notes that feedback may

Figure 2.2
How Companies Use 360-Degree Assessment Results

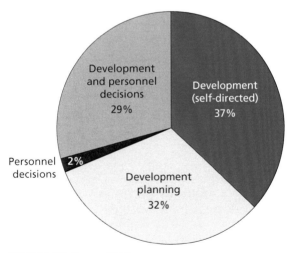

SOURCE: 3D Group, 2013.
RAND RR998-2.2

---

[2] The results of the 3D Group study appear to be based on a convenience sample of organizations. As a result, the findings are not necessarily representative of all organizations in the United States. Nevertheless, the sample does include a large number of well-known corporations. At a minimum, the study can provide useful insights into the ways in which those organizations use and view 360s.

be inflated because raters find it stressful and threatening to share negative comments, raters may have a vested interest in the ratee receiving a favorable result, and ratees may nominate raters who are more likely to provide positive comments. Experts also note that anonymity can enable raters to sabotage the ratee or can lead the ratee to believe that raters intend to sabotage him or her, which in turn will lead the ratee to discount negative feedback. While there are few studies that directly compare 360-feedback systems used for development versus evaluation, most experts still advise against using them for evaluation.

Employee Views on 360 Feedback

There is evidence that employees generally have favorable reactions to 360s. For example, responses to surveys of more than 1,400 employees in multiple companies (Edwards, Ewen, and Vendantam, 2001) revealed that over 75 percent of respondents reported that the program added useful information and improved process fairness, felt that feedback was helpful, and recommended the 360 process. Data from 59 project evaluations conducted in the late 1990s across a range of industries show even more favorable reactions to 360 programs (Edwards, Ewen, and Vendantam, 2001).[3]

Participants' reactions to 360s are important because they affect ratees' beliefs about the quality of the feedback and their willingness to use it. In the literature on performance evaluation more generally, studies show that reactions, such as satisfaction with feedback and perceived feedback accuracy, influence subsequent performance (Jawahar, 2010). Research on 360 feedback showed that individuals who received less-favorable feedback had more-negative reactions and perceived the feedback as less accurate; those who had more-negative reactions and found the feedback less accurate reported that the feedback was less useful (Brett and Atwater, 2001).[4] (For a comprehensive review on performance feedback in the workplace, see London, 2003.)

Studies have also identified a number of factors that influence reactions to 360s, and many of these results point to strategies for 360-feedback system design and implementation. One factor is the purpose of the feedback; studies have shown that participants prefer 360s for development rather than for evaluation. Participants believe that positive outcomes are more likely and negative outcomes are less likely when using subordinate and peer feedback for development as opposed to evaluation (e.g., Bettenhausen and Fedor, 1997; Farh, Cannella, and Bedeian, 1991; Fedor, Bettenhausen, and Davis, 1999). For example, Fedor,

---

[3]   In the latter study, consisting of the 59 project evaluations, data were available regarding the purpose of the evaluation, i.e., for evaluation or development. The authors reported responses for only one item—the usefulness of the results—and, contrary to expectations, found higher agreement rates in companies using the appraisal for evaluative rather than developmental purposes. However, the authors also note that this result could have occurred by chance.

[4]   These findings should not be interpreted to mean that raters should avoid providing unfavorable feedback. In a study of leaders in an elite unit of the U.S. military, Smither, London, and Richmond (2005) found that individuals who received more unfavorable ratings (and who were perceived by a psychologist as more defensive) set more goals for improvement six months later than individuals who received more positive ratings (and who reported the feedback as more valuable). In addition, in a study of performance evaluation, Jawahar (2010) found that characteristics of the rater and the way in which feedback was formulated and delivered influenced ratee satisfaction, which in turn affected use of the feedback; ratees had more favorable reactions when they viewed the rater as credible, the feedback was based on job criteria, and the rater refrained from criticism and engaged in goal setting with the ratee. These findings have implications for design and delivery of performance feedback.

Bettenhausen, and Davis (1999) found that nurses were more positive toward peer reviews when they perceived the reviews to be developmental rather than evaluative, and Farh, Cannella, and Bedeian (1991) found that, although participants had similar opinions about the value and accuracy of ratings, they were more likely to recommend peer feedback for developmental than evaluative purposes.

Evidence of Performance Improvement Using 360s

360s are popular in large part because participants (raters, ratees, and managers) generally like them. But liking 360s is not the ideal measure of success when they cost an organization significant time and resources. As shown in Table 2.1, success of 360s can be defined in a number of ways. However, from the organization's perspective, behavior change should be the ultimate goal. As such, Bracken, Timmreck, Fleenor, and Summers (2001) believe that 360s are successful when they create "focused, sustained behavior change and/or skill development in a sufficient number of people so as to result in increased organization effectiveness" (p. 4).

Acceptance of the process by the raters and ratees can be an important precursor to behavior change in that it can impact the usefulness, accuracy, and acceptance of the feedback itself. As a result, the other measures of success should not be disregarded. Nevertheless, failing to explore whether meaningful and desirable behavior change is occurring could lead an organization to waste significant resources on something with little improvement in organizational effectiveness.

Research shows that 360 feedback results in small but positive improvements in individual performance. A meta-analysis of 24 studies found that, on average, managers' performance improved following 360 feedback; however, the magnitude of the improvement was small (Smither, London, and Reilly, 2005). Findings were similar regardless of the role of the evaluator (i.e., subordinate, peer, supervisor, or self), whether the person being evaluated received feedback only from subordinates or from subordinates and others (i.e., peers and/or supervisors), and the type of study design used. "The most likely explanation for the small effect sizes is that some feedback recipients are more likely to change than others" (Smither, London, and Reilly, 2005, p. 45). This is consistent with the results of a larger meta-analysis of feedback interventions more generally (i.e., not 360 feedback exclusively), which found that, while most individuals improved, about one-third showed decreases in performance (Kluger and DeNisi, 1996). In light of their review findings, Smither, London, Vasilopoulos, et al. (1995) stated the following:

> [I]t is unrealistic for practitioners to expect large across-the-board performance improvement after people receive multisource feedback. Instead, it appears that some feedback recipients will be more likely to improve than others. We therefore think it is time for researchers and practitioners to ask "Under what conditions and for whom is multisource feedback likely to be beneficial?" (rather than asking "Does multisource feedback work?"). (p. 60)

A few studies have also shown inconclusive results with respect to predicting overall organizational performance. For example, Pfau and Cohen (2003) found that, while higher-quality human resource management practices were positively associated with firm financial perfor-

Table 2.1
Examples of How 360 Success Could Be Defined from Various Viewpoints

Constituent Type	Definitions of Success	Factors Contributing to Success
Rater	• Perception that feedback was accepted and ratee behavior improved • Improved working relationship • No negative repercussions • Improved feedback climate	• Anonymity • Ratee accountability • Multiple administrations • Policies and practices communicated • Hotline to report problems
Ratee	• Received feedback that was fair, constructive, valid, actionable, credible, and reliable • Improved communications • Improved working relationship • Improved feedback climate • Improved performance	• Consistency • Confidentiality • Rater training • Rater accountability (i.e., honesty) • Consistent rater selection • Sufficient number of raters with knowledge (i.e., census) • Valid, reliable instrument • Organization commitment (e.g., resources) • Access to raters for follow-up • Development and action planning
Supervisor	• Access to timely, quality feedback appropriate for use in performance management and/or development • Feedback aligned with organization goals and/or values (i.e., relevant, important) • Improvement in ratee behavior consistent with the process' objectives • Improvement in ratee workgroup • Improved feedback climate • Not overly intrusive (time, cost)	• Supervisor receives data that are clear, actionable, relevant • Supervisor receives training on proper use of data • Ratee accountability for follow-through • Organization commitment (e.g., resources) • Process is a business priority • Process is administered regularly, consistently • All supervisors held accountable • Valid, reliable instrument • Efficient data collection, reporting
Organization	• Sustained focused behavioral change in large number of individuals that leads to organizational effectiveness • Viability of the process (sustainable) • Improved feedback climate • Legal defensibility • Supports learning environment	• Top management commitment • System-wide implementation • Resources to develop valid process • Sufficient resources to sustain the process • Valid, reliable instrument • Alignment • Efficient, cost-effective data collection, processing, and reporting

SOURCE: Bracken, Timmreck, Fleenor, and Summers, 2001.

mance, 360-feedback programs were associated with a 10.6-percent decrease in shareholder value. The authors attributed this finding to ineffective implementation of 360-feedback programs. However, the results were correlational; it is possible that companies that suffer from a decrease in shareholder value were more likely to implement 360-feedback programs. In addition, shareholder value is only one measure of organizational performance, and it is not a measure that is particularly relevant to military organizations.

As noted by many researchers (see, for example, Bracken and Rose, 2011), however, few 360 efforts ever include measures of success.

Designing and Implementing a 360-Feedback System

Once an organization decides that 360 feedback is appropriate, numerous design and implementation factors need to be considered. Some of these factors include the following:

- Is the organization ready for a 360-feedback program? Is 360 feedback consistent with the organizational culture and goals?
- Will 360 feedback be used for development only, for administrative decisions only, or both?
- Who should be involved in the system design and implementation?
- What content should be included on a 360-rating form?
- What types of ratings should be used?
- Should the form include open-ended questions?
- Should participation for both ratees and raters be voluntary or compulsory?
- Who chooses the raters?
- How many raters of each type are needed (e.g., supervisors, peers, subordinates, customers)? How many raters should be invited from each source?
- How often will feedback be collected?
- How will results be presented to ratees (e.g., aggregated within source and/or across sources)? Presented along with normative data? Presented over time?
- Who sees the results?
- Should raters be anonymous or identified?
- How will participants be trained?

Unfortunately, there are no universal best answers to each of these questions. Instead, each decision needs to be tailored to the organization's unique context and needs. As such, there can be quite a bit of variability in the implementation of 360s across contexts and organizations. Table 2.2, which summarizes the results of the 3D Group's (2013) survey of 360-feedback practices in 211 U.S. and Canadian companies, illustrates some of that variability.

Table 2.2
Summary of Company Experiences Using 360 Feedback

Which types of 360 programs does your organization use?		Who developed the survey and report format?	
Individual: The 360-degree feedback process is made available on an as-needed basis.	78%	Committee/team (internal)	33%
Training and development: The 360 program is part of a larger training and/or development program.	45%	Consultant	22%
Enterprise-wide: All or nearly all of the managers within the organization participate.	23%	Internal committee or team with a consultant	20%
Department-wide: Specific departments or units go through the 360 process.	17%	Bought "off the shelf"	20%
		Other	5%
How many points are on your rating scale?		**Does your rating scale focus on:**	
5	87%	Agreement (such as "strongly agree" to "strongly disagree")	49%
4	6%	Effectiveness (such as "not effective" to "very effective")	31%
6	2%	Frequency (such as "always" or "daily")	23%
7	2%	Importance (such as "very important" for the job)	17%
Other (3, 9, or 10)	3%	Other	7%

Table 2.2—Continued

How many questions does your survey contain?	
11 to 40	53%
41 to 69	21%
Up to 10	17%
70 to 100	6%
Over 100	3%

How many open-ended questions are included?	
Two	22%
Five	20%
Three	16%
Zero	15%
Six to ten	11%
One	7%
Four	7%
Over ten	3%

How, if at all, do you present variation in scores across raters?	
Illustrate graphically	37%
Raw scores—frequency distribution	33%
None	14%
Standard deviation	13%

What standard measure, if any, do you have to which feedback recipients can compare their results?	
Organization norms	34%
No standard measure	30%
Previous years' scores	26%
National norms	19%

Who chooses raters?	
Participant with manager approval	42%
Human resources (based on organizational chart)	25%
Participant only	17%
Manager only	10%

How do you train your raters to complete surveys?	
Test included in survey	56%
Web training	19%
In-person training	13%
Email	6%
Teleconference	6%

Who receives a copy of the report?	
Participant	96%
Human resources	60%
Participant's direct manager	59%
360 coach	24%

What, if any, is the minimum number of raters per rater category (e.g., peers, direct reports) for data to show in a report?	
Three	38%
No minimum	19%
Two	11%
Four	10%
Five	10%
Six	7%
More than six	5%

SOURCE: 3D Group, 2013.

Fortunately, however, there is an enormous amount of theoretical and empirical research addressing the details associated with the design and implementation of 360s that can help guide organizations in making many of these decisions. Because covering all of the relevant issues is far beyond the scope of this report, we direct organizations to consult the existing literature about best practices in 360 feedback to guide their efforts. Sources do vary, however, in the extent to which they provide sound, research-based information and recommendations. *The Handbook of Multisource Feedback* (Bracken, Timmreck, and Church, 2001) is a particularly comprehensive and useful resource to support implementation; it consists of chapters addressing a wide range of topics, written by leading experts in 360-feedback research and practice. We direct interested readers to consult that resource as a starting point.

CHAPTER THREE

Performance Evaluation and Promotion Processes in the Military Services

Performance evaluation in the military has traditionally consisted of top-down evaluations in which supervisors rate their subordinates. With this top-down approach, there is no opportunity for subordinates to contribute content directly and systematically to performance evaluation. The current performance evaluation system also does not provide a formal opportunity for review of a leader's performance by peers of similar rank. In an organization as hierarchical as the military, this could mean that important facets of leadership and managerial performance are omitted from performance evaluation information. A 360-degree assessment is one method that some think would be useful to help fill in those gaps. It is a tool that the military services use to varying degrees, which we discuss in the following chapter. As context for considering how or whether 360s might be used more broadly as a component of performance evaluation and promotion processes in the military services, we begin in this chapter with a description of those systems.[1]

Performance Evaluation Systems

Officer performance evaluation reports are governed primarily by regulations issued by the individual services. Each service has developed its own philosophy, policies, and procedures for evaluating the performance and potential of its officers. There are, however, three provisions of law applicable to all services that prescribe specific action regarding performance evaluations. The first requirement—found in Title 10 of the United States Code; Part II, *Defense Acquisition Positions*; Chapter 87, *Defense Acquisition Workforce*; Section 1722, *Career Development*—addresses performance appraisals reviewed by a person serving in the same acquisition career field.[2] The other two provisions impose a requirement for the chairman of the Joint Chiefs of

---

[1] This overview of current performance evaluation and promotion processes in the military services is based primarily on a document review of laws, policies, and procedures that was verified and expanded during interviews with representatives from each of the services.

[2] This section requires the secretary of each military department to "provide an opportunity for review and inclusion of any comments on any appraisal of the performance of a person serving in an acquisition position by a person serving in an acquisition position in the same acquisition career field." This section is implemented by the Under Secretary of Defense for Acquisition, Technology, and Logistics (Department of Defense [DoD] Instruction 5000.66, 2005).

Staff to assess the performance of an officer as a member of the Joint Staff or other joint duty assignments.[3]

While each service prescribes its own procedures and criteria for evaluating the performance and potential of its officers, there is some commonality among the services:

- *Evaluation reports.* Three services prepare reports for O-1s through O-8s; the Army prepares reports only for O-1s through O-7s.
- *Annual reporting cycle.* Most reports are prepared annually, but the services have procedures to accommodate unique situations that may affect this annual cycle.
- *Top-down evaluations.* The person who prepares the assessment and subsequent reviews is in the supervisory chain above the officer being evaluated. Peers and subordinates have no input into the preparation of performance evaluations.
- *Rating chain.* Three services use multiple raters; the Navy uses a single rating official.
- *Service advisors.* Advisors are designated to assist rating officials when the officer being evaluated is assigned to an organization or command outside the officer's service and the rating chain does not include an officer in the same service as the officer being evaluated.
- *Evaluation criteria.* Each service has established its own set of competencies and attributes by which it measures an officer's performance and potential to assume greater responsibility. Leadership is the only evaluation criterion common to all services.
- *Written comments.* Rating officials provide written comments on the performance of the officer being evaluated.
- *Assessment metric.* A rating scale is used for officers in certain grades, with those grades determined by service policy.
- *Promotion recommendation.* Three services include an assessment of an officer's potential to assume greater responsibility; the Air Force accomplishes this function on a separate form.
- *Signature.* The officer being evaluated must sign the performance report after all reviews are complete.
- *Appeal process.* Each service has a multistep process that allows an officer to raise concerns about an evaluation. The final avenue for redress is an application to the service board for correction of military or naval records.
- *Counseling.* Each service includes individual counseling as part of the performance evaluation cycle. Three services include counseling as an integral part of the evaluation process; the Marine Corps views counseling as distinct and separate, but complementary to the performance evaluation system.

Table 3.1 provides an overview of the similarities and differences of the major elements of performance evaluation systems used by each of the military services.

---

[3]   The first of these two provisions calls for the chairman of the Joint Chiefs to provide the assessment when an officer has been nominated to a position of importance and responsibility, designated by the president, which carries the grade of general, admiral, lieutenant general, or vice admiral (Section 601 of Title 10, United States Code). The provision calls for an assessment when an officer has been nominated as a commander of a combatant command; commander, U.S. Forces, Korea; or deputy commander, U.S. European Command, but only if the commander of that command is also the Supreme Allied Commander, Europe (Section 604 of Title 10, United States Code). Under the second provision, the chairman may also include consideration of other aspects of the officer's performance as the chairman considers appropriate.

Table 3.1
Comparison of the Military Services' Performance Evaluation Systems

Service	Evaluation Form	Rating Chain	Evaluates Performance	Evaluates Potential	Evaluation Approach	Appeal Process	Counseling[a]
Army	• Company grade • Field grade • Strategic grade[b] • Academic	• Rater • Intermediate rater • Senior rater	Yes	Yes	Rating scale (company and field grades)[b] Narrative	Yes	Yes
Navy	• Ensign to captain • Rear admiral	• Rating official	Yes	Yes	Rating scale Narrative	Yes	Yes
Marine Corps	• Lieutenant to major general[c]	• Reporting senior • Reviewing officer • Third officer sighter	Yes	Yes	Rating scale (O-5 and below) Narrative[d]	Yes	Separately[e]
Air Force	• Lieutenant to colonel • Brigadier to major general • Academic	• Rater • Additional rater • Reviewer	Yes	In the narrative[f]	"Does not meet standards"[g] (O-6 and below) Narrative	Yes	Yes

SOURCES: Army Regulation 623-3, 2014; Chief of Naval Personnel, 2011; Commandant of the Marine Corps, 2010; Air Force Instruction 36-2406, 2005; and representatives who provided input in our interviews.

[a] Frequency and method of counseling vary among the services. The Army and Air Force use separate forms to facilitate counseling and require counseling to be conducted face to face if at all possible. The Navy leaves the counseling method to the commanding officer or officer in charge.

[b] In the Army, Air Force, and Marine Corps, officers in the pay grades of O-1 to O-3 are referred to as company grade, O-4 to O-6 are field grade, and O-7 and higher are strategic grade (general officers). The equivalent officer groupings in the Navy are called junior grade, midgrade, and flag.

[c] Marine Corps generals and colonels: Only specified information in the administrative section is completed; evaluation is submitted using letter format.

[d] Marine Corps colonels: Consideration and evaluation of attributes are addressed in the narrative letter.

[e] Marine Corps: Counseling is a separate but complementary process to the performance evaluation system.

[f] Air Force O-6 and below: Uses a separate process specifically to make a promotion recommendation.

[g] The Air Force system provides for an evaluation in several criteria whereby an officer is rated as "meets" or "does not meet" standards in each area.

Tables 3.2 and 3.3 compare the evaluation criteria established by each of the military services to evaluate its officers. To help illustrate some of the similarities and differences, we sorted each service's evaluation criteria into three conceptual categories (professional or technical expertise, personal qualities, and leadership and command climate).[4] The words in bold type are the criteria, and the description following the titles provides more detail on what the rater actually evaluates. Because some criteria are relatively self-explanatory, a description is not provided for all criteria, but each service has a more detailed description of the attributes for each criterion in its governing regulation. Table 3.2 shows the criteria for O-6s and below; Table 3.3 for general and flag officers.

While each service has its own approach to and criteria for assessing performance, they all achieve similar results. The system establishes the performance characteristics that are important to the service and documents how well the officer met those standards. That information

---

[4]   Our category groupings are not necessarily reflective of the services' groupings of these criteria.

Table 3.2
O-6 and Below Performance Evaluation Criteria

Service	Professional/ Technical Expertise	Personal Qualities	Leadership/ Command Climate
Army	• **Achieves:** gets results, prioritizes limited resources to accomplish the mission, accomplishes mission consistently	• **Character:** adheres to Army values, empathy, warrior ethos/service ethos, discipline; supports SHARP, EEO/EO • **Presence:** military bearing, fitness, confidence, resilience, professional bearing • **Intellect:** mental agility, sound judgment, innovation, interpersonal tact, expertise	• **Leads:** leads others, leads by example, builds trust, extends influence beyond chain of command, communicates • **Develops:** creates a positive command/workplace environment, prepares self, develops others, stewards the profession
Navy	• **Professional expertise:** knowledge, proficiency, qualifications • **Mission accomplishment and initiative** • **Tactical performance** (warfare-qualified officers only)	• **Military bearing/character:** appearance, conduct, physical fitness, adherence to Navy core values	• **Teamwork** • **Leadership:** organizes, motivates, develops others to achieve goals • **Command climate** • **Equal opportunity**
Marine Corps	• **Mission accomplishment:** performance, proficiency • **Fulfillment of evaluation responsibilities**	• **Individual character:** effectiveness under stress, courage, initiative • **Intellect and wisdom:** decisionmaking ability, judgment	• **Leadership:** leading/developing subordinates, ensuring well-being of subordinates, setting the example, communication skills
Air Force	• **Job knowledge** • **Organizational skills:** plans, coordinates, schedules, and uses resources effectively	• **Judgment and decisions** • **Physical fitness** • **Professional qualities:** loyalty, honesty, discipline, dedication, integrity, officership	• **Leadership skills:** fair and consistent, fosters teamwork, sets/enforces standards, motivates subordinates, promotes healthy organizational climate • **Communication skills**

SOURCES: Army Regulation 623-3, 2014; Chief of Naval Personnel, 2011; Commandant of the Marine Corps, 2010; Air Force Instruction 36-2406, 2005; and representatives who provided input to our interviews.

NOTES: SHARP = Sexual Harassment/Assault Response and Prevention Program; EEO = equal employment opportunity; EO = equal opportunity.

is used in turn for a variety of personnel actions, most notably during selection board deliberations in which promotions decisions are made.

Promotion Process

Unlike performance evaluation systems, the process for selecting and promoting officers is prescribed in law.[5] DoD issues implementation-related policies, and each service publishes regulations specifying how to comply with the law and implement DoD policies. The laws governing officer promotion in the military services fall primarily under Chapters 35 and 36 of Title 10, United States Code. Chapter 36 contains most of the promotion process provisions; Subchapter I prescribes the requirements for selection boards (when they are to be convened,

---

[5]   The promotion laws described in this section are relevant to active-duty officers. Additional chapters of Title 10 of the United States Code cover promotions for reserve officers.

Table 3.3
General/Flag Officer Performance Evaluation Criteria

Service	Professional/ Technical Expertise	Personal Qualities	Leadership/ Command Climate
Army	• **Achieves:** accomplishes the mission, maintains/builds organization's capabilities	• **Character:** adheres to Army values, warrior ethos, moral and ethical qualities • **Presence:** projects a commanding presence • **Intellect:** mental agility, sound judgment, innovation	• **Leads:** leads by example, builds trust, extends influence beyond chain of command • **Develops:** creates a positive command climate, seeks self-improvement, steward of the profession
Navy	• **Fiscal planning/ organizational skills** • **Professional growth** • **Mission accomplishment** • **Operational/professional competence**	• **Vision/strategic thinking** • **Geopolitical fluency** • **Military bearing** • **Personal growth**	• **Communication skills** • **Leadership/judgment** • **Leading change**
Marine Corps	• None specified		
Air Force	• None specified • Comment on professional and personal characteristics		

SOURCES: Army Regulation 623-3, 2014; Chief of Naval Personnel, 2011; Commandant of the Marine Corps, 2010; Air Force Instruction 36-2406, 2005; and representatives who provided input to our interviews.

composition of the board, information to be furnished to the selection board, reporting and approval requirements), and Subchapter II deals with promotions (criteria a selection board must consider and details of the promotion process, such as eligibility for promotion). Two sections of Chapter 35 prescribe the procedures for appointment of officers to certain three- and four-star positions. Section 662 of Chapter 38 prescribes minimum promotion objectives for officers serving or who have served on the Joint Staff or have been designated as joint qualified. In addition, sections in other chapters of Title 10 contain specific requirements with which promotion selection boards must comply when considering officers for promotion.

OSD and Service Guidance

The Secretary of Defense has issued regulations to implement the provisions of law regarding selection boards and the promotion process. The two primary instructions are DoD Instruction 1320.14, *Commissioned Officer Promotion Program Procedures* (2013) and DoD Instruction 1320.13, *Commissioned Officer Promotion Reports* (2014). Each of the services has issued implementing instructions regarding the promotion process as prescribed in Title 10 and DoD policy. Of particular note, each service policy specifies the information provided to selection boards as prescribed in Section 615 of Title 10:

- Army Regulation 600-8-29, *Officer Promotions* (2005), states that promotion boards are to receive the "performance portion of the OMPF [Official Military Personnel File]." Army Regulation 623-3, *Evaluation Reporting System* (2014), specifies that evaluations will be maintained in the soldier's Official Military Personnel File.
- Secretary of the Navy Instruction 1420.1B, *Promotion, Special Selection, Selective Early Retirement, and Selective Early Removal Boards for Commissioned Officers of the Navy and Marine Corps* (Secretary of the Navy, 2006), states that the Chief of Naval Personnel and

the commandant of the Marine Corps "shall supply all pertinent records of each officer to be considered by the board. Such records shall include all documents, including fitness reports. . . ."

- Air Force Instruction 36-2501, *Officer Promotion and Selective Continuation* (2014), specifies that the name and officer selection record is information given to the board. Air Force Instruction 36-2406, *Officer and Enlisted Evaluation Systems* (2005), states that an evaluation becomes a matter of record once it has been filed in the officer selection record.

These requirements bear directly on decisions concerning the use of 360-degree feedback in the military services. Specifically, if 360s become a formal part of the evaluation system and, thus, are included in officer performance evaluation reports, the results of the 360s would, under current law, be included in the information provided to selection boards. Whether this is a desirable practice should be considered before making a decision to use 360s for evaluation purposes in the military, as is discussed later in this report.

Summary

This review demonstrates that each service sets its own performance evaluation policies, following standards that best meet its respective needs. That the services tailor their performance evaluation systems is an important attribute to keep in mind when considering whether and how to further implement 360s in DoD. For example, should the department decide to mandate the use of 360s, even in a limited way, such as for select leadership positions, it may be advisable to allow the services to determine an implementation approach that aligns with their current evaluation processes.

In contrast, the law controls the promotion process. Numerous provisions establish the requirements for determining eligibility, establishing and conducting a selection board, and specifying the process that must be followed for an officer to be promoted. Perhaps the most important attribute of the promotion process as pertains to the use of 360s is the fact that performance evaluations are a formal part of an officer's selection record and thereby provided to selection boards for consideration. Nothing in law precludes the use of a 360 or the inclusion of 360s in promotion considerations. However, should 360s be mandated as part of performance evaluations, they would necessarily be provided to selection boards unless current laws are modified. This too is an important consideration in determining the best use of 360s within the department.

The chapters that follow review current use of 360s in the military services, discuss various uses of 360s, and evaluate these alternatives within the military context.

CHAPTER FOUR

360-Feedback Systems Currently in Use in the Military

All of the services have channels for feedback to personnel outside of the traditional top-down review systems. These can be formal, such as the Marine Corps' "Requesting Mast" (a Marine can speak privately about his or her concerns with someone several ranks higher); informal, such as casual conversation and discussions across ranks and peers; event-specific, such as after-action reports in the Army and the Marine Corps (where debriefs are conducted after a mission); or environment focused, such as command climate surveys, which focus on the qualitative features of command, such as integration and honesty. 360s are currently just one more structure for aggregating feedback from outside the direct command chain.

Each service and the Joint Staff are using some version of a 360. Table 4.1 summarizes several features of the 360 programs we identified in our military stakeholder interviews.[1] As shown in the table, there are large areas of commonality in the utilization of 360s across the services. For example, each service is using 360s in some form for development, and none has 360s in place for performance evaluation. However, there are some clear differences as well. For example, the Army, with the most experience by far, has spent 14 years developing its MSAF program, which now has widespread implementation. The Army is also the only service that requires that each officer complete a 360.

Other services, such as the Marine Corps, which is just starting a pilot, have far less experience, and their programs tend to be smaller and narrowly targeted. Most are directed either at senior leadership or toward high-potential officers as part of the military education system. A number of the programs are in early stages of development and are being piloted on small subsets of the services' respective populations. All of the programs are developmental in nature, and their small scale typically allows for personalized coaching support in the interpretation and utilization of the results. Despite the shared developmental focus, there is variation across these programs in how widely the results of the assessments are shared. In many cases, only the individual and his or her coach—if he or she elects to have one—see the results. However, in both the Air Force and the Joint Staff, the chief and chairman (respectively) can access general officers' 360 results.

Given that the Army has the largest existing 360 effort and the NDAA specifically points to it as an exemplar, the Army's programs are described in detail below. Following that, we describe some other existing efforts in the services that could serve similar purposes to a 360.

---

[1] Although this is a fairly comprehensive list of the formalized 360 efforts in the services, it is likely that there are other efforts scattered throughout the services (such as those embedded in larger training programs or that are taking place in other informal development settings) that are not captured in the table.

Table 4.1
Examples of Existing 360 Efforts in Each Service

Service	Tools Used	For Whom?	Who Provides Ratings?	Is It Required, and How Often?	Who Sees the Results?
Air Force	Air Force Leadership Mirror	Many Air Force populations	Participant selects (subordinates, peers, and superiors)	On taking grade	Individual
	Air Force General Officer 360	General officers	Any general officer, command chief master sergeants, colonels, GS-15s, Senior Executive Service members	Yes, annually	Individual, chief and vice chief of the Air Force
	CCL Executive Dimensions	Lieutenant generals and Tier-3 Senior Executive Service	Participant selects	Taken at CCL at career milestone	Individual with the option of coaching
	CCL Benchmarks	Select colonels, GS-15, E-9	Participant selects	Taken at CCL at career milestone	Individual with coaching
Army	Leader 360 (MSAF)	All officers (O-6 and below), enlisted and Department of the Army civilians	Participant selects (CAC holders)	Yes, once every three years and tracked with checkbox on Officer Evaluation Report	Individual with the option of coaching
	Commander 360 (MSAF)	Battalion and brigade commanders on the Centralized Selection List	Rater selects, commander can nominate two for inclusion	Yes, three to six months and 15–18 months into command	Commander feedback report, must be reviewed with rater (who is trained to provide coaching)
	Peer and Advisory Assessment Program	General officers	Anyone in general officer corps who feels capable of providing feedback	Yes, annually	Individual with the option of coaching, Army GOMO reviews results for red flags
	General Officer 360	Those promoted to two-, three-, and four-star posts	Participant selects with Army GOMO approval	When promoted to next general officer post, approx. every three years	Individual with the option for coaching, Army GOMO reviews results for red flags
Navy	360 Assessment at Navy Leadership and Ethics Center	Officers screened for command	Participant selects	On command selection	Individual with one-on-one feedback from certified coach at the training
	Surface Community Pilot 360 Program	Department heads	Participant selects	Early career (about seven years of service)	Individual with debriefing by a certified feedback coach at Surface Warfare Officer Department Head School
	New Flag Officer and Senior Executive Symposium	New flag officers	Participant selects	A week-long training course	Debriefed by executive coaches

Table 4.1—Continued

Service	Tools Used	For Whom?	Who Provides Ratings?	Is It Required, and How Often?	Who Sees the Results?
	Naval Post-Graduate School Flag Officer 360	Rear admiral (upper half) and vice admiral	Participant selects	While attending a customized course specifically geared toward preparing them for their next assignment	Debriefed by executive coaches
	CCL Flag Officer 360	Vice admirals and admirals	Participant selects	At the Leadership at the Peak course	Debriefed by executive coaches
Marine Corps	Center for Army Leadership MSAF 360 (Pilot)	Colonel/lieutenant colonel commanders	Participant selects (CAC holders)	No, pilot (MSAF) run once at Commandant's Commanders Program	Individual with the option of coaching
	CCL Executive Dimensions	Lieutenant generals and Tier-3 Senior Executive Service	Participant selects	Taken at CCL at career milestone during the Leadership at the Peak course	Individual with the option of coaching
Joint	Joint 360	Joint Staff and combatant command general and flag officers	Participant selects (subordinates, peers, and superiors)	Yes, six to nine months after taking post; again in two years	Chairman, deputy chair, and director J7 may request to see an individual Participant also receives report and may choose self-review, review with coach, supervisor, or a senior mentor

NOTES: Table is populated based on interviews conducted by RAND researchers and supplemented with data provided by the office of the Senior Advisor to the Secretary of Defense for Military Professionalism in OSD. GS = General Schedule pay scale; CCL = Center for Creative Leadership; CAC = Common Access Card; GOMO = General Officer Management Office.

The Army's Multi-Source Assessment and Feedback Program

The Army currently operates the most expansive 360 program, known as the Multi-Source Assessment and Feedback (MSAF) program. The MSAF is intended to enhance a leader's adaptability and self-awareness, to identify strengths and weaknesses, and to prepare the individual for future leadership responsibilities. The assessment gathers feedback from multiple sources—peers, subordinates, and superiors—using questions that focus on core leader competencies and supporting leadership behaviors as defined in Army leadership doctrine (Army Doctrine Reference Publication 6-22, 2012). The results are used to establish self-development goals and self-improvement courses of action, and participating leaders have the option of gaining assistance in this process from a coach. The Army governing regulation states that this program is separate and distinct from formal performance evaluations and personnel management practices. Although results are not made available to the rating chain, whether an individual completed the MSAF and the date of the most recent assessment are noted in annual performance evaluations (i.e., the Officer Evaluation Report) for lieutenants through colonels, although the results are accessible only by the individual service member.

Every officer through O-6, noncommissioned officer, and all Department of the Army civilians must complete a 360 review at least every three years. There are multiple versions of the tool and multiple channels for its administration. Every version is developmental in nature, reinforces the same core principles as outlined by Army leadership doctrine, and is integrated into a central hub—the Virtual Improvement Center. The current program has over a decade of implementation history and development and continues to evolve today. Within the past two years, the Army has restructured its standard instrument and broadened its technical scope. This process is continuing to expand the acceptance of 360s and foster a culture of feedback for the Army.

MSAF 360 Implementation

The MSAF program is available to any service member or Department of the Army civilian with a CAC. Each individual is tasked with logging in to the system; completing a self-assessment; and inviting peers, subordinates, and superiors to provide feedback. According to the Center for Army Leadership, completing the online instrument takes between ten and 12 minutes. Prior to beginning the process, individuals are encouraged to take online training modules that outline the purpose and importance of the MSAF, the process of utilizing the MSAF for self-development, and training for raters on how to accurately assess an officer's behavior. The minimum number of required raters is three superiors, five peers, and five subordinates, though participants are informed that the quality of the assessment result is improved with increased number of assessors. Assessors can be drawn from across the armed services, and the Army is increasingly streamlining the process to gather feedback from multiple branches.

After the survey is completed, the individual being reviewed receives an individual feedback report that contains average ratings from each rater group—peers, subordinates, superiors, and others—as well as the free-form comments. The individual reports are maintained exclusively for the eyes of the user. The MSAF portal, which provides the tool and houses the report, also contains links to development materials, such as videos, doctrines, and further self-assessments housed in the Virtual Improvement Center. The Army's objective is to continue to improve the development center so that feedback results are immediately linked to relevant training, videos, or policy documents.

In addition, all individuals are encouraged to seek coaching and mentorship. The Army provides access to full-time coaches—professional outsiders, as well as retired personnel trained for the task—and encourages the rated individual to discuss the results with his or her superior. The Army currently has the fiscal resources to provide 1,500 coaching sessions annually, which matches demand—approximately 1 percent of eligible Army leaders seek out coaching, and just shy of 1,500 coaching sessions are used annually. An evaluation of the coaching service revealed that 90 percent of the participants were satisfied with their coaching experience, and 30 percent sought more than one coaching session.

Use and Impact of MSAF Results

The current MSAF program, rolled out in 2008, has experienced a continuous increase in use of approximately 10 percent per month. As of September 2014, 335,000 assessments had been conducted under the program. Data from a 2013 Army evaluation of the MSAF program indicated that close to one-third of participants list personal self-development as the primary motivation for completing a 360.[2] Of those who complete the survey, approximately one-half seek follow-on discussions with others on the results. The most common form of follow-on discussion is with peers and colleagues (37 percent), followed by discussions with a supervisor (28 percent).

We are not aware of any published evidence that captures the efficacy of the Army's 360 efforts in terms of effects on behavior or performance; however, published studies and unpublished data provided by the Center for Army Leadership provide some insights on participants' views about them. In a 2013 program evaluation, Army leaders indicated that they find the program useful and insightful for their development. Approximately one-half of the participants said that the process revealed an unknown aspect of themselves, 33 percent said they discovered a weakness, and 17 percent said they discovered a strength. Two-thirds of users indicated intent to change their behavior as a result of the survey results, though only around 10 percent of assessors said that they observed behavior change in individuals who had gone through 360s. And two-thirds of the surveyed population either agreed or strongly agreed that the MSAF provides valuable insight.

On the other hand, other data suggest that, while some might find 360s useful, many may not perceive them to be as effective as other leadership development activities. For example, in the Center for Army Leadership's annual survey of leaders (Riley et al., 2014), only about one-quarter of participants rated the MSAF as having "large or great impact" on their development as leaders, another quarter rated the MSAF as having a "moderate impact" on their development as leaders, and 50 percent indicated that it had "small or no impact." Out of the 15 leadership development practices about which the survey asked participants, the MSAF was among those with the lowest perceived impact;[3] however, as noted by researchers at the Center for Army Leadership, the MSAF is the least expensive and least time-consuming assessment practice. Both sources suggest that there are some individuals who find it beneficial.

The most prevalent challenge that MSAF users face is obtaining an adequate number of individuals to complete the online assessment. According to the 2013 MSAF evaluation survey, 70 percent of users found getting enough respondents for their MSAF to be the biggest challenge. Currently, there is no system in place to compel individuals to complete feedback forms for others, and there is no system for ensuring that certain individuals do not get inundated with large numbers of requests.[4] This problem has been increasing as the MSAF has increased in use.

MSAF Costs

The Army prides itself on the ability to execute this program in a cost-efficient manner and has developed the experience to maintain and improve the tool in house. Recent improvements have included a redesign of the questionnaire—shortening the questionnaire by 40 percent and validating the new instrument—and adding the capability to incorporate email addresses

---

[2] Data on the use of MSAF results were provided to RAND during interviews with personnel at the Center for Army Leadership.

[3] Ranking based on postsurvey analysis. Participants were not asked to provide a relative ranking, only an assessment of each practice independently.

[4] Completing assessments of others is required by Army Regulation 350-1 (2014). However, there is no system to force completion.

across the entire DoD to increase the usability of the system by non-Army personnel (mostly as reviewers).

The current program uses eight contract-year manpower equivalents (CMEs)[5] for operational support and information technology support, which is the largest cost. An additional four CMEs are devoted to coaching. There is one full-time program manager, a GS-13, two Army majors who serve full-time as action officers, and a GS-14 lead scientist (1 CME). The resources to develop and pilot the program are not included in these costs and were not available to our research team.

Other 360 Efforts in the Army

The Army has formally implemented at least three other variants of the 360 tailored to address specific audiences, contexts, and goals.

The first is the Unit 360, which is the same instrument as the MSAF but, at a unit commander's request, is administered to a whole unit concurrently. For commanders who request a Unit 360, the Army provides support tools, such as a program manager who can ensure that the instrument is properly distributed, track the process of completion, and answer any questions. The manager has no access to the content provided in the feedback forms. The individual results of a unit event are identical in all respects to the MSAF and are kept strictly confidential. However, when an entire unit participates in such an assessment, a unit roll-up is prepared that includes aggregate scores across a unit, summary reports, and recommendations. The unit, the unit commander, and the commander's superior officer see the results, which can be used to guide training and development efforts for the unit.

The Army also offers a peer evaluation version of the MSAF for the general officer corps, in which general officers provide feedback on any officer of whom they have sufficient knowledge to rate. This feedback is meant to be developmental, but it is reviewed internally for any indications of misconduct or toxic leadership—instances have occurred in which peer feedback motivated internal scrutiny and, ultimately, disciplinary action. The Army has also recently implemented a 360 evaluation exclusive to general officers, which includes colonels as well as other general officers as raters.

In October 2014, the Army added the Commander 360 program, which replaces the MSAF requirement for lieutenant colonels and colonel-level commanders in the Centralized Selection List. The Commander 360 is completed twice during the course of command assignment, once within three to six months of assuming command and again after 15 to 18 months of command. The Commander 360 is similar to the MSAF in that it uses the same technology, is based on the same leadership doctrine, and is kept separate from the formal Officer Evaluation Reports, but the questions are tailored to address the specific demands and challenges of the command role.

There are three other differences between the MSAF and the Commander 360: (1) The commander's superior (i.e., rater) identifies the peers, subordinates, and superiors who will

---

[5]   Salaries for contractors were not available to our team, but a U.S. Office of Personnel Management white paper found contractors to be cost saving relative to equivalent-level government employees. In 2012, the approximate base salary for a GS-9 was $48,000, for a GS-11 was $58,000, for a GS-12 was $70,000, for a GS-13 was $83,000, and for a GS-14 was $99,000 (U.S. Office of Personnel Management, undated).

provide feedback; (2) the commander and the rater are provided the report; and (3) the commander and his or her rating superior must engage in a developmental discussion following the receipt of the feedback report to plan for the commander's growth as a leader. The superiors are instructed not to use the information contained in the 360 for evaluation. The Army relies on the professionalism of the senior staff to consider the feedback appropriately, as well as understand the context of the results. The Commander 360 program is designed to encourage candid and informed discussions about leadership and development between commanders and their superiors. The desired goal is an active developmental plan for the officer. The Commander 360 program was pilot-tested in July 2013 and April 2014 and received favorable reviews from participating commanders and their raters.

The Command Climate Survey is another tool in the Army designed to provide leaders with views from below. Although command climate surveys are used by all of the services, in the Army it is a supporting piece of a larger feedback effort. The Command Climate Survey captures participants' perspectives on many aspects of command (including unit climate, morale, and absence or presence of sexual harassment and equal opportunity violations). Most questions do not focus on an individual but rather on resources, trust, and cooperation. The influence of the commanding officer is reflected in these answers but few questions pertain to his or her performance directly. The results of these surveys are publicly shared with the unit, the commanding officer, and the commanding officer's superior officer. These surveys are relatively short, can be initiated on short notice, and are not viewed as a replacement for the 360.

Related Efforts

During our interviews, participants identified other tools both in the Army and the other services that might produce similar insights to that of 360s—command climate surveys, Inspector General reports, the Marine Corps' Request Mast, and informal information collection. Although their purposes might be broadly similar to those of 360s, they do not solicit the same range of feedback in a systematic way. Whether these tools could be duplicative of the 360 ultimately depends on how they are structured and implemented.

Command Climate Surveys

All the services administer climate surveys that commanders periodically use to assess the working environment within a service unit and/or command. We discuss the Army's use of two such command climate surveys above. Three interviewees described command climate surveys as duplicative of the 360 review, while four described them as playing a complementary role.

Regarding the duplicative role, for example, the Marine Corps' command climate survey includes questions intended to solicit subordinates' perception of their commanding officers. Example closed-ended questions ask respondents to provide agreement with statements such as, "My CO [commanding officer] makes clear what behavior is acceptable and not acceptable in my unit;" "My CO clearly communicates priorities for the unit;" "My CO establishes an environment that promotes mutual trust and respect." The Marine Corps' climate survey also asks open-ended questions, including (1) "What is your unit doing well, that it should continue to do?" (2) "What is it not doing well, that it should improve upon?" and (3) "What do you know about your unit that leadership does not know, but should?" The commanding officer can, but is not required to, share the results with his or her senior commanding officer or rater, as well as his or her team. According to our interviews, commanders typically view the process as developmental and are open to sharing results with the senior raters and subordinates.

The Air Force's mandatory Defense Equal Opportunity Survey, a recent update of the Unit Climate Assessment, also asks questions about the environment and people that could be considered duplicative of a 360 review. According to an interviewee, "it is not done in a 360 way but in a way that brings out stuff that [is] similar to what a 360 would. . . . It is administered anonymously at each level [and] focuses on things like sexual assault–related environmental questions."

Similarly, according to our discussions with Navy personnel, the 360 may be duplicative of some elements of existing Navy command climate surveys. The Navy representatives with whom we met expressed a preference for the command climate survey over the 360 and indicated that the command climate program is already useful for identifying and weeding out toxic leaders. As one participant explained, "I am sure every senior is using that [command climate survey] to color their evaluation. . . . It is the basis for firings sometimes." According to descriptions of the Navy's current command climate assessment questionnaire (Navy Personnel Command, 2014; Ripple, undated), it aims to assess several aspects of organizational effectiveness and equal opportunity, such as fair treatment, sexual assault prevention and response, favoritism, diversity management, organizational processes, intention to stay, and job burnout (Ripple, undated). Focus groups are also conducted as part of the process (Navy Personnel Command, 2014), and example questions include, "What do you like most and least about your command?" "How difficult is it for you to recover from making a mistake in the eyes of your supervisor?" and "In a word, how would you describe your command?"

Although some participants in each of the services suggested there could be overlap between the 360s and the climate assessments, the extent of the overlap would depend entirely on the content and design features of both the 360 and the command climate assessment. That is, they could be designed in such a way as to capture similar or different aspects of performance. Duplication of such information could be useful in some contexts; however, to the extent that unintentional duplication exists, it would be wasteful.

Inspector General Reports

All military personnel have the right to file complaints with the Office of the Inspector General. According to one interviewee, every year,

> there are thousands of online complaints that tend towards ethical, misconduct, and legal infraction of which I think 20 percent gets substantiated, and, of those, only another 20 percent are punitive in nature. There is also a DoD hotline where officers can call at any time to make complaints about toxic leadership and other ethical concerns.

The point is that subordinates already have an avenue through which to surface toxic leadership, and that avenue is actively used.

Marine Corps Request Mast

Every Marine is entitled to participate in the Request Mast procedure, whereby he or she can ask to speak with a superior commander, commanding general, or inspector general about an issue. In issuing the request, the officer does not have to disclose the issue he or she wants to discuss, which allows officers the opportunity to anonymously express concerns to senior leadership.

Informal Sensing

Interviewees from all the services agree that there are informal ways of soliciting subordinate input and that good senior leaders already do so. Yet they disagreed on whether informal solicitation and feedback gained through 360 reviews are duplicative. One interviewee felt that they might be duplicative because "there are no secrets on a ship, so I don't know that such a survey tool used for the purposes of evaluation is necessary to do that." This interviewee further supported this view by saying that if a supervisor is unable to do this, then the supervisor should be held accountable for his or her subordinates' actions.

In comparison, another interviewee shared the perspective that formalizing the feedback makes the difference: "Theoretically, there is nothing in there that supervisors couldn't get by just going out and talking to your subordinates, except that it is now formal and on paper." Another expressed the similar view, that

> walking around in an organization, you can get a sense of its health, but 360 is a way to get more granularities. It is an early-detection tool. We don't use 360s to detect criminal behavior, we want to use 360 as an early-detection tool, get a sense of the organization or personal qualities or characteristics [of the leader]."

CHAPTER FIVE

Using 360 Feedback: Evaluation Versus Development

This chapter and the next explore answers to the two remaining research questions posed in Chapter One: *Would it be advisable to implement 360-degree assessment for developmental or evaluation purposes in the military?* And *If DoD were to implement 360s, what are the implementation challenges that the services should be aware of to ensure success?* Although each chapter focuses on a separate research question, the answers to both questions are inextricably intertwined. When we presented these questions to our DoD participants, they often discussed implementation challenges at the same time as they debated whether 360s should be used for development versus evaluation purposes.

Nevertheless, we chose to present these two questions independently, beginning in this chapter with a discussion of the issues most hotly debated by our interviewees in responding to the question of whether 360s should be used for development versus evaluation purposes. In Chapter Six, we present other implementation concerns raised by our participants about the use of 360s in the military—many of which apply to the use of 360s for either development or evaluation. Each chapter includes a summary of the insights gained during our interviews and a discussion of key points of overlap between the DoD perspectives and the extant research literature and best-practice approaches to 360s.

DoD Perspectives on the Use of 360s for Evaluation

Do military stakeholders believe that it is advisable to implement 360s for developmental or evaluation purposes in the military? The short answer is "no" for evaluation and "yes" for development.

Participants clearly see many more negatives than positives when it comes to using 360s for evaluation purposes, as shown in Table 5.1. Moreover, the negative sentiments were spread fairly evenly across all of the service groups with which we met (see Table 5.2). Even representatives from the Army, which is already administering the MSAF to all officers for developmental purposes, cited many reasons for not implementing the tool for evaluation.

Those who were supportive of using 360s for evaluative purposes generally focused on using them for service members in leadership positions. Proponents of the use of 360s for evaluation believe that subordinate input is essential to selecting effective leaders. One even went so far as to say that selection through evaluative use of the 360 may be of greater value than its development use, though both are valuable.

Participants offered up numerous explanations for why they believe that the negatives outweighed the positives in using 360s for evaluation purposes. Among the most strongly

Table 5.1
Comments About Using 360s for Evaluation Purposes

Negative Comments		Positive Comments	
Incorrect data or dependent on who responds	15	Might help with toxic leadership	7
Rater dishonesty problems	12	Might improve leadership	3
General concern about use of 360 data to inform promotions	12	Might help with ethics violations or misconduct	4
Concern that there might be incorrect or harmful use of results	12	Might improve performance	1
There would be too much cultural resistance	11		
Concern that using for evaluative purposes will ruin developmental use	11		
Administrative and time cost is wasteful	9		
"Popularity contest" conflicts with mission objectives	9		
Would cause legal issues for promotions	7		
Financial cost	6		
Might duplicate existing efforts (i.e., climate surveys)	6		
Need further research to determine efficacy and cost/benefit	2		

NOTE: Throughout the interviews, participants described various pros and cons and problems and obstacles to 360s for evaluation purposes. This table illustrates the range of the negative and positive comments provided. Each interview is counted as only one comment regardless of how many people were participating in the interview or how many times the idea was repeated.

argued reasons were that they could cause leaders to be unwilling to discipline their subordinates and that the raters would no longer be willing to be honest. But perhaps the biggest concern with using 360s for evaluation purposes is that they could ruin the value of 360s for development. All of these are discussed further in the next chapter.

If 360s Were Used for Evaluation, Certain Ways Would Be More Advisable

Even the interviewees who supported using 360s for evaluation still suggested that they be used only under certain circumstances and conditions. Others who were opposed to their use for evaluation suggested that, if they were to be implemented, certain issues would need to be clearly addressed. The sections that follow provide greater detail on these issues.

Using 360s to Inform Leadership Selection Decisions

Although our participants overwhelmingly oppose using 360s for evaluation, some did acknowledge that having a way to evaluate someone's performance from a range of viewpoints would be useful in making command or leadership selection decisions.

Table 5.2
Negative Comments About Evaluative Use, by Service Group

Comment	Air Force	Army	Marine Corps	Navy	Joint
Incorrect data or dependent on who responds	X	X		X	X
Rater dishonesty problems	X	X	X		X
General concern about use of 360 data to inform promotions	X	X	X	X	X
Concern that there might be incorrect or harmful use of results	X	X	X		X
There would be too much cultural resistance	X	X	X	X	X
Concern that using for evaluative purposes will ruin developmental use	X	X	X		X
Administrative and time cost	X	X	X	X	X
"Popularity contest" conflicts with mission objectives	X		X	X	X
Would cause legal issues for promotions		X	X		X
Financial cost	X		X	X	X
Might duplicate existing efforts (i.e., climate surveys)		X	X	X	X
Need for further research to determine efficacy and cost/benefit			X	X	

They believed that collecting data that could provide a more complete view of candidates' performance and making such data available would be beneficial when one reaches a certain level of command:

> The only way we eliminate narcissism from senior executives is if we pick senior executives that have some insight into how they behave towards their subordinates. Most of the tools that we currently use for selecting and assigning people are devoid of measures of the impact that leaders have on subordinates; how they generate enthusiasm, trust, and respect; and how they care for their subordinates. This is such a delicate area that I am certain that we need multiple assessments for selection.

One suggested that it could be useful to include 360s as part of a larger commander evaluation suite that could include assessment centers (which employ a series of leadership and workplace simulation activities), command climate surveys, and the Army's Commanders 360. Another suggested that it could be useful for everyone in a command position, including company command, battalion, and brigade to platoon leaders. Other proponents recommended their use only above certain ranks (such as O-5/O-6) because of the overall potential for impact, resource constraints, and too much movement among subordinates below those ranks.

Even though there were a few proponents, most of them still felt that 360s would be useful only if they were implemented in certain ways and under certain conditions. One said, if used for evaluative purposes, 360s should

> explicitly ask questions that are geared towards identifying problems. The surveys should not have questions where you rate someone on a scale of 1 to 5 and hope that somewhere in there you will identify a trend. I think we need to ask people—to the tune of, "In the last year, have you observed your commander violating any ethical rules that you are aware of?"

Another participant suggested that if information could be retained over the course of a career, perhaps it could be shared at the point when command and general officer selection decisions were being made. Another suggested that if subordinate feedback were solicited for selection boards, it should not be called 360 so as to distance it from the developmental value of the process.

As an alternative, another person suggested a process similar to the background interviews used for security clearance investigations—a process in which trained interviewers interview past subordinates and peers confidentially.[1] The data collected during these interviews could be considered when evaluating the leadership and character of potential senior leaders prior to selection to high-level positions (command, colonel, or general officer level), thus offering a confidential bottom-up look for selection decisions. Choosing this approach for selection would be less likely to taint the value of the 360 process for development purposes, because the format would be entirely different. It also might create an environment in which subordinates would be comfortable providing honest input and allow the interviewer to provide the much-needed context for such information, thus addressing two major concerns with using 360s for evaluation purposes.

360-Feedback Reports Could Be Used to Inform Supervisor Ratings

Permitting 360 information to be shared with a supervisor and allowing that supervisor to review the information and decide whether and how to incorporate it into someone's performance evaluation was explored in some of our interviews. This seems to be the most benign way of incorporating 360s into evaluations, and probably the only way that key stakeholders in the services would not reject the idea outright. However, most participants were still strongly opposed to this use, again because of concerns about it ruining the use of 360 tools for development.

Several participants pointed out that good supervisors already seek feedback informally from subordinates for someone's formal evaluation, so gaining information via 360 feedback would be no different. On the other hand, 360 provides a more structured and granular assessment that is not available to supervisors who are seeking this information informally. A hybrid approach, in which feedback is provided to the supervisor and the ratee long before the evaluation period, is another approach that allows the supervisor and ratee to set goals for improvement together. During the ratee's formal evaluation, the supervisor could then evaluate whether the ratee met those goals. If the subordinate made improvements by the time that evaluations were conducted, that should reflect positively on the ratee in performance evaluations. If improvements were not made, the supervisor could take that into consideration as well.

---

[1]   For more information about this proposed alternative, see Taylor, 2014.

Context Would Need to Be Provided If Used by Promotion Boards

Even those who support the use of 360s for evaluation purposes agree that contextual understanding is necessary to properly interpret 360 results. Supervisors would likely have to be consulted to help provide that context. For example, in some jobs, there are few subordinates. Some assignments are tough assignments in which subordinates are going to dislike anyone in a command position. In some cases, subordinates will have little opportunity to observe someone's performance. In other cases, they may see only one side of it. In some jobs, the position may be extremely challenging. In others, it could be relatively unchallenging. Sometimes, a peer or subordinate might have a personal agenda or provide information that might be entirely false. Additionally, the type of leadership needed changes as service members move from position to position and climb the ranks. In a lower-level position, someone's performance might be viewed as adequate for that level, but at higher levels, it might not.

Given the myriad of context differences that could occur, even large differences in 360 ratings about two different individuals could be meaningless. Without some way to account for the context, the score differences could be easily misunderstood. Several participants commented about just this issue:

> There is a balance [that needs to be achieved] between giving everybody an opportunity to provide feedback and getting information that's not correct. For developmental, this is not a threat. You can give the information back to the person and to [his or her] supervisor to explain or make sense of it. But for evaluations it would be hard. . . . You are just getting raw data.

> One of the benefits of being in the chain of command when you evaluate somebody's 360 is that you have context for the information. If something you find raises a red flag and you don't have context, you may draw the wrong conclusion from the comments.

> The hard ratings that the ratee receives for doing what the service needed them to do would not accurately capture the context. Some people come in and keep a ship on course, and some people come in and have to shift things. This person is at a disadvantage and maybe this story doesn't get captured when you are just looking at the numbers.

As noted by participants, if implemented for evaluation and used by promotion boards, such contextual differences would have to be explained. That would likely require significant input from a knowledgeable supervisor, and supervisors would need to be well trained in how to provide that context.[2]

---

[2] Although supervisor ratings might also suffer from such context issues, the implicit assumption is that supervisors are better equipped to understand the context than subordinates because they have a broader view of the organization as a whole and performance of the individual than others have. Given that, in the military, subordinates are typically exposed to only a few superiors, whereas a supervisor has a view of many subordinates for comparison, it seems reasonable to conclude that supervisors would be better equipped to provide the needed context in an individual's performance evaluation. Nevertheless, the recognition by many participants that even the supervisor would need training in how to provide the needed context is an acknowledgment that supervisor input itself can vary in quality and accuracy.

DoD Perspectives on the Use of 360s for Development

In contrast to the largely negative comments about using 360s for evaluation purposes, interviewees expressed many more positive comments about developmental use of 360s (Table 5.3). They also expressed some concerns (also shown in Table 5.3), but the issues were fewer than those expressed about evaluative use. Additionally, participants reacted far more positively to the idea of considering 360s for developmental purposes, even in light of the negatives they mentioned.

A number of those interviewed talked about why 360s have value when used for development. They believe that 360s help identify and promote positive traits, as well as highlight weaknesses, which gives officers an opportunity to strengthen those areas. In general, officers want to understand their strengths and weaknesses, and 360s are an opportunity to gain insight—a tool that can facilitate "conversation between leaders and subordinates about what is needed to develop."

> When I did it, I sought out people who will have nothing to gain by telling me I was a great guy. I wanted to find out where I could grow. You take the survey yourself and other people grade you. In my experience, I graded myself harder than everybody else. And in the military, that tends to be common. In the commercial world, it is the opposite. Your typical military guy wants to grow. The temptation to whitewash things may happen in the [evaluation] side but not in the development side.

> [We get] tremendous feedback on the ones we do. Everyone treasures it. I would say for 85 percent who do a 360, it's the first one they ever had. And when I use words like, "Hey,

Table 5.3
Comments About Using 360s for Developmental Purposes

Negative Comments		Positive Comments	
Administrative and time cost	8	Might improve leadership	13
Need for further research to determine efficacy and cost/benefit	7	Might help with toxic leadership	12
Concern that superiors will use developmental results in an evaluative way	7	Might improve performance	10
Concern that there might be incorrect or harmful use of results	7	Raters might be honest	8
Incorrect data or dependent on who responds	5	Might help with ethics violations	7
Financial cost	5	Might help with misconduct	7
Rater dishonesty problems	3	The service would be culturally receptive to it	7
There would be too much cultural resistance	3		
"Popularity contest" conflicts with mission objectives	3		

NOTE: Throughout the interviews, participants described various pros and cons and problems and obstacles to 360s for developmental purposes. This table illustrates the range of the negative and positive comments provided. Each comment is counted only once per interview, regardless of how many people were participating in the interview or how many times the idea was repeated in the interview.

did it help you identify some blind spots or gaps?" or "Wow, I didn't really know I came across that way," . . . they all [say], "You bet; that's exactly what happened."

Proponents of 360 assessments believe that officers can grow from an approach that provides different perspectives on how attitudes and behaviors are perceived by others:

> It helps people to see gaps that they have. We may all perceive that we all have certain leadership traits and personalities that are more effective than they actually are. They may speak to a certain group but not to another group. So [the feedback] helps you focus on things that [you] do or say that I think is very open but ends up threatening people. It helps you see how you could be a more effective leader, certainly on the developmental side.

> The character piece is a blind spot for folks [that the 360 can address].

Unlike evaluative use, there were some differences across services in the number of negatives that were raised in using 360s for developmental use. Interviewees from the Army tended to express the fewest negative comments about using the 360 for development, and the greatest number of positive comments, although they too recognized that there are limits to its usefulness. The Air Force and the Joint Staff also had many positive comments about its use for development. The Navy and Marine Corps representatives had positive things to say as well, although they were the most reserved in their support for use of the tool.

Overall Conclusions on Advisability of Evaluation Versus Development

From the summary of the comments provided above, it is clear that most stakeholders believe that the disadvantages of using 360s for evaluation purposes outweigh the advantages. Even the Army—the organization that cited the fewest downsides to developmental 360s—cites many reasons it could be detrimental and ill-advised as an evaluation tool. Although these results alone would suggest that our participants would argue against using 360s for evaluation in the military, we also asked participants directly whether they would be supportive of using 360s for each purpose (development or evaluation). A breakdown of their answers is shown in Table 5.4.

Stakeholders were overwhelmingly against using the tool for evaluation. Fourteen out of the 19 interview groups provided us with an emphatic "no" when asked whether using 360s for evaluations was advisable. Although most participants provided us with a clear yes or no answer when asked directly, two did not. One explained that developmental uses should be in

Table 5.4
Number of Interview Groups That Believe It Is Advisable to Use 360s for Evaluation Versus Development

Response	Development	Evaluation
Yes	16	3
It depends	3	0
No clear answer	0	2
No	0	14

place first before the services even consider exploring using 360s for other purposes. The other stated, "It would take some time and study," and either way, "the purposes [developmental and evaluation] should never be merged. If you mix the two, it will be problematic." Three interview groups responded "yes," indicating that they would consider using 360s in evaluations but that many safeguards would need to be in place—that 360s should be used only for certain groups and/or under certain circumstances and that more research would be needed to explore consequences of their use over time. Some of the conditions under which they could envision using 360s for evaluation purposes are described in the next section.

In contrast, 16 of the 19 interview groups expressed clear support for using developmental 360s for a subset of the officer population. The remaining three provided slightly more cautious responses. Two of these recommended that they be used for development but only on a voluntary basis. Of those two, one recommended that they be used only with more information about the cost–benefit ratio before a full-scale rollout, and the other felt that "if it is voluntary, then the people [who] least need it will do it and the ones [who] need it [won't] do it." Of the 16 groups that clearly recommended using 360s for developmental purposes, 12 recommended them only for those at the O-4 or O-5 ranks and higher. Most cited resource constraints as the rationale for limiting 360 use to those in higher grades. This topic is covered in greater detail in the following chapter. Only two interviewees, both from the Army, recommend using the 360 for development for all military personnel, one describing it as "totally worth it at all levels," and the other as being useful for "everyone, with openness between ratee, superior, and subordinate."

Not Ideal for Addressing Toxic Leadership, Ethics Violations, and Other Problem Behaviors

In addition to asking participants whether 360s should be used for evaluation or for development purposes, we specifically asked participants whether they thought 360s would be useful in addressing toxic leadership or other problem behaviors—whether as a development or evaluation tool. A majority felt that eliminating toxic leadership was an important goal but that 360s were not the best tool for accomplishing it. Instead, they pointed to other tools already in place that are specifically designed to address that need:

> There are all sorts of methods—open-door policy, chaplain, observation, complaining, . . . etc. It is a brigade commander's responsibility to know whether his [or her] battalion commanders are toxic. By observing and talking to soldiers two levels down, you can find out what you need to know about the quality of the leaders under you.

> There are already tools in place . . . [such as the inspector general process] to find out the environment. Toxic leaders are being found this way.

> There's also mandatory assessments, such as the Defense Equal Opportunity [Climate] Survey . . . which asks about the environment and people. It is not done in a 360 way but in a way that brings out stuff. It is administered anonymously at each level. It focuses on things like sexual assault–related environmental questions.

Although they did not feel that addressing toxic leadership was the main purpose of 360s, participants did note that some toxic leaders might be improved with 360s. Participants discussed three types of toxic leaders. The first is the toxic leader who means well but is unaware of how his or her actions are viewed by others or is unable to improve on his or her own. For those toxic leaders, 360s for developmental purposes might open their eyes to their actions and ultimately have a positive impact on their behavior. The second type is the toxic leader who has no desire to change. Participants agreed that many toxic leaders are of this ilk, and, for them, 360s for developmental purposes would likely have no effect. They also noted that, if 360s were included in evaluations, they might have an impact on these people (because evaluations are high-stakes), but these individuals are so rare that using 360s in this way would be an expensive and time-consuming way to find them. Participants again pointed to other ways of finding these people that would be much more cost-effective (such as through anonymous reporting channels, climate surveys, informal discussion, or inspector general complaints). The third type of toxic leader described is generally viewed negatively because he or she makes tough decisions that subordinates dislike. This individual might appear toxic on paper, but, in actuality, he or she would be doing exactly what was most needed. Participants believed that 360s might backfire in such cases as this.

Ethics violations and other problem behaviors were generally discussed as a different issue. For those types of behaviors, the belief was that (1) reporting of that kind of behavior would lead to a myriad of legal due-process concerns; (2) ratees would be no more likely to report it on a 360 than they would using the other avenues available to them, and they should be reporting it through those avenues; and (3) most people already know when someone is engaged in problem behaviors—a 360 would not provide any new information. The latter belief—that everyone already knows—was also expressed about toxic leadership:

> Most people have worked for a toxic leader at some point, and it wasn't a secret. Everybody knew who was a jerk. The question is: What was done about it? You could find them; then what? What did we do about it? Doesn't the culture determine what happens at that point? Do we need 360 or do we know? And the bigger question is what are we doing about it? In many cases, alert leaders know, but would they do something about it?

Finally, respondents suggested that if the goal of using 360s for evaluation was to find toxic leaders or bad behavior, 360s would create a policing culture that would not only undermine the benefits of 360s but also engender greater paranoia and distrust in the system:

> The 360 is not a magic wand. It is just a tool for leader development that could have a positive impact, but it may not. Take, for example, the guy who had an issue at Fort Bragg. I don't know if he had a 360 or not and, if he did, it may or may not have had an impact on the behavior he displayed. To say we want to do it to mitigate over-the-top behavior of some senior officials is not the angle we are coming at it from. When people are senior, over 50 years, they are the way they are, a 360 is not likely to change them several folds. The bottom line is that we view this as a developmental tool far more than [as] a tool to prevent behavior that a small percentage [of service members] exhibit.

CHAPTER SIX

Using 360 Feedback: Other Implementation Issues

This chapter addresses the second of the two implementation questions: *If DoD were to implement 360s, what are implementation challenges of which the services should be aware to ensure success?* Whether the services use 360s for development, evaluation, or some combination of the two, numerous issues would need to be ironed out when implementing this tool. Some are of a practical nature, such as identifying the right raters, coaching, and legal issues. Other concerns are more esoteric in nature, yet equally important—such as how to use 360 feedback without eroding trust and how to ensure acceptance within the military culture. System fundamentals, such as the accuracy of the ratings and the logistics required for implementation, also need to be considered. All of these topics were raised during the course of our interviews, and many are also echoed by experts in the field.

Identifying the Right Raters

As noted in an earlier chapter, in practice, some 360 feedback tools do not involve all three traditional feedback sources: subordinates, peers, and supervisors. Although some experts in 360-degree assessment (see, for example, Foster and Law, 2006) argue that feedback tools that exclude one or more of the three sources or that include other sources (such as customer ratings) should technically be called "multisource feedback" rather than 360, we included these other forms of multisource feedback in our broad definition of 360 for the purposes of this review. Our participants also conceived of it broadly, offering numerous thoughts about who might be most appropriate as raters in military 360s.

Overall, participants' views differed on who should participate in 360s; but despite the differences in perspective, the general goals of their comments were similar: to gather the right information in an efficient way and without undue duplication. Several people with whom we spoke believed that a full 360 review with peers, subordinates, and supervisors all providing feedback would not be necessary or even beneficial in the services. Instead, they suggested that a partial 360 using only some of the sources would be better. Some felt that peers and subordinates might be interchangeable and that little additional insight would be gained from spending the resources on capturing information from both. Others felt that peers and subordinates had unique perspectives. Still others argued that the contribution of peers versus subordinates would depend entirely on the rater's and ratee's positions.

Several people suggested that the main insight of the 360 is the subordinate perspective, and so greater emphasis should be placed on that source:

> For boards, for a variety of reasons (administrative simplicity is one but not the main one), I would . . . use only subordinate input. Subordinate input would not be [from] the people [who] are immediately working with this person; they could be included, but you should go back to prior commands and go back to prior people. There are a number of advantages in that, to include enhancing the anonymity of the process.

Others felt that the supervisor already provides structured feedback and so tasking the supervisor with completing a 360 is wasteful, because any information gathered is likely to be duplicative of the standard performance evaluation.

Some felt that, in the typical military workplace setting, peers have little opportunity to observe each other's performance and hence would have little to contribute. In some military training courses, however, peers may have plenty of opportunity to observe someone's performance and may be quite well situated to provide appropriate feedback. Others argued that peer feedback should be avoided at higher levels of command, because the climate can become competitive and affected by politics. The competitive nature of these positions would reduce the value of feedback provided by peers such that the exercise would not be worth the investment:

> Peer input can be useful. But peer input is particularly sensitive in competitive places. And senior people, it's a very delicate thing and I don't think they add that much. And the potential for friction and competition and lack of opportunity to observe—an awful lot of O-6s in the services don't really see how other people are doing.

Still others believe that a full 360 (with supervisors, peers, and subordinates) is beneficial in certain circumstances:

> Yes, go for the full range in development. The boss is already there, but you want self-compared to boss, peer, subordinate. Hopefully the boss is evaluating and giving feedback frequently, but yes, for development purposes we need the true 360.

Moreover, as noted by a participant, even a rater's familiarity with the ratee can be important:

> Another challenge is that the assessed officer will send it to someone who doesn't really know [him or her]. If we are going to use it for evaluation, whether we are asking the right people to do the assessment is critical.

Experts generally agree that the usefulness of the sources can vary from setting to setting, and even within one setting, subordinates or peers may be able to comment about only a subset of performance. One way to maximize the usefulness of the information is to tailor the questions to each source (DeNisi and Kluger, 2000).

The point to be taken away from these varying perspectives is that part of establishing a 360-feedback system is to give careful consideration to who should participate. In part, the answer lies in the purpose for the 360 and how the service intends to use the results—a topic discussed in the following chapter—and in part the answer lies in the type of position.

Incorporating Coaching and Follow-Up

Many experts in 360s recommend incorporating coaching or some other form of follow-up activity to help improve the likelihood that 360s will have positive impacts (see, for example, Atwater, Brett, and Charles, 2007; and DeNisi and Kluger, 2000). Indeed, there is evidence from empirical research that coaching can have benefits for subsequent job performance. In one study, managers who met with direct reports to discuss their upward feedback improved more than managers who did not, and they improved more in years when they discussed the previous year's feedback than in years when they did not (Walker and Smither, 1999). Reviewing plans and progress quarterly, seeking input from coworkers, and receiving coaching and feedback are strongly related to improvement in ratings over time (although managers engaged in these activities relatively infrequently) (Hazucha, Hezlett, and Schneider, 1993). Another study found that systematic coaching on how to interpret discrepancies between self-ratings and other ratings improved manager self-awareness and subsequent performance (as rated by others), as well as attitudes and commitment of both the managers and the other employees (Luthans and Peterson, 2003). However, experts acknowledge that coaching itself can be tricky; thus, the people providing the coaching need to be skilled in it. This means that managers or coaches need to be trained in how to interpret the feedback and how best to deliver it, another added cost to implementing 360s properly.

The importance of including coaching was emphasized repeatedly in our interviews. There was near-universal agreement among those who have had experience with 360s that properly structured coaching is essential for successful implementation. In many cases, this opinion was shaped by experience with the smaller 360 programs found throughout the services that target senior officers. In a summation of a common sentiment about coaching, one interviewee said, "We think the coaching will be the most critical piece for getting developmental utility."

Participants also brought up coaching issues to consider, including who should provide the coaching, when and with what frequency, and the feasibility of providing coaching to large numbers of personnel. The value of extended conversations with coaches who are able to interpret the data and assist with development plans was mentioned as an important element for success:

> The people I have talked to and myself when I've had 360s, I find the great value came from having a great coach outside of the organization, who takes [these] data and is trained in how to assess [them]. And sits down in a very lengthy conversation with me, often an hour, walking through the results, asking those probing questions that in the end I say, "hmm, I have some blind spots; here [are] some areas that I can maybe improve in."

> How you follow up and create a plan is magic. There needs to be some sort of plan and commitment to a continuation. It can't be a one-off. It's a developmental tool. What are you going to do about this information? What do you do to strengthen the response?

> Anyone [who] uses a coach finds it more rewarding, and that's the most expensive part. And you have to get that right.

Coaches are often the largest expense of the programs and are either retired senior officers or professional outside coaches. The Army, which has the largest 360 program, has internalized the coaching function:

> The expectation is that the rater will sit down and mentor the officer. This is not the single source for coaching. Just like with any other MSAF, all officers still have access to professional coaches if they want, and they can click on access to the doctrine and virtual support. Training is being developed to help the raters learn how to be . . . better [mentors].

Several participants also brought up the theme of integrating the feedback into service culture. One person brought up that integrating feedback evaluation into the military education curriculum is valuable to create comfort with and commitment to the system. Another person spoke of the broader cultural change that feedback can engender:

> I believe that we have to move towards important meaningful dialogues about "how am I doing?" Not just "how am I doing with getting the job done?" which is important and happens now. But "how am I doing with building camaraderie and cohesion and building professional relationships?" which requires some coaching and mentoring and some frank discussions with some officers . . . .

Addressing Fairness and Legal Concerns

Participants were concerned that a number of fairness and legal concerns might arise as a result of collecting 360 information. Most were associated with either the fact that a rater's responses are anonymous or decisions regarding when something would trigger an official investigation.

Interviewees' concerns about anonymity were more notable in the evaluative context. When 360 ratings are anonymous or confidential (that is, either raters cannot be identified or their identity is otherwise protected), a number of potential legal implications arise if confidentiality concerns impact the formal evaluation process (either officially or unofficially). If a ratee ultimately disagrees with information in a 360 but it goes into his or her official record, how can the ratee appeal? In a traditional fitness report, there is due process, but for 360s appeals and official investigations, this might not be workable if confidentiality is to be maintained. As one interviewee put it,

> There is nothing to go back and prove them wrong. They'll never know who submitted the information. With the appeals process, you can get negative information removed from your promotions record. If there is something bad there, somebody will come back to say "we have to prove this thing." . . . In the end, the bad thing hardly gets reported because the anonymity gets lost in investigation.

> Our system of government gives you a right to offer a rebuttal. How do you do that in an anonymous confidential survey where the people filing it want to remain anonymous?

Several interviewees explained that anonymity might be more difficult to protect when using 360s for evaluation purposes because military personnel have the right to learn about adverse information that affects their performance reviews and promotion. Some expressed concerns that it could lead to investigations intended to identify the anonymous rater. One stated,

> It is harder for us to keep that confidential if someone is asking for it, if it is adverse and being used for evaluation. If for developmental, then as long we can show that it didn't leak into the evaluations, it's not a problem.

With respect to anonymity, a common fairness concern among interview groups was whether anonymity could actually be preserved when the pool of possible raters is narrow:

> If I supervise 15 people or five people and I have to do a 360, when someone calls out something, I know exactly who said it.

There was also concern over the exploitation of anonymity. Interview participants worried that an individual could use anonymity to communicate strong sentiments without regard for validity or accuracy of those assessments. Potential misuses of the anonymity of the 360 span concerns over incomplete data, biased data, and collusion and sabotage:

> It is difficult to get people to do the survey because, no matter what they said, people were concerned about the anonymity.

Another concern raised by our interviewees was the appropriate course of action if something illegal or that otherwise violates military codes of conduct is raised in a 360. For example, if a complaint is filed against an officer, typically an investigation occurs to determine the accuracy of the information. But, if the information is guaranteed to be anonymous or confidential, is the organization even legally allowed to take action on it? As an interviewee questioned,

> Who, if anyone, is supposed to check for such information? Do we keep quiet because it is anonymous? Do we do an investigation or not? I'm not sure. . . .

Maintaining Trust Within the Organization

Discussions of 360s often mention trust in several ways. Some described general concerns that 360s would lead to distrust of one's peers and subordinates and the organization. The armed services see trust as an essential feature of mission preparedness, and the impact of the 360 is a high concern. In our interviews, we found a range of opinions on this topic. The greater portion of our participants was concerned that 360s—particularly for evaluation purposes— could erode trust that is needed for units to operate successfully. There is concern about what 360s "might do to the cultures of teamwork and competence [and about] what the impact will be on those." There are also concerns that 360s might be perceived as negatively affecting service members' careers—and in doing so undermine trust in the broadest sense:

> A key concern is the impact on organizational trust: How do you foster cohesion in unit? Our command climate tells us we could do a better job of building trust. An assessment 360 has the potential to cause people to cover up rather than open up. We depend on trust, and anything that breaks that down is not good.

Right now the services are downsizing; putting in one more component that could under-mine a career will increase the paranoia about whether a service member is going to have a job tomorrow.

Others felt that the tool could be effectively used in higher ranks without issue but that service members at more-junior grades of lieutenant and sergeant might not have the maturity to handle the tool without adversely affecting trust:

> I think the possibility of creating an environment where day-to-day mutual trust and respect is somewhat compromised by this . . . when we are talking about people at that level. I have used, and I think it's good to use, these instruments at the very basic level if it is done in a proper environment. But basically I am not sure that company commanders and [their] lieutenants and sergeants and so forth are quite ready to handle it. This is my secondary concern.

And still others felt that the 360 can foster greater transparency and lead to an overall improvement in trust:

> They could be useful for establishing greater trust. The Army has a unit 360, and I'm really impressed with the results. It strikes me as really sharp. It starts from the developmental perspective and can improve transparency in the unit.

Another topic raised by participants and experts alike (for example, DeNisi and Kluger, 2000) related to the importance of trust associated with the confidentiality of the feedback itself. Without confidentiality of the results, raters might not be honest, and ratees might not be motivated to use the information in constructive ways:

> Confidentiality rules will make or break the developmental effort. The subordinates are not going to give you and me accurate, candid feedback if they think [they] are going to be known. It's got to be sure that the boss does not know who we are.

> Confidentiality and debriefs by certified coaches are the centerpieces of the development approach. This confidential feedback (1) creates a supportive and trusting environment for the growth and development of its officers; (2) promotes confidence in the results; (3) ensures those providing responses retain their confidentiality without fear of attribu-tion; and (4) provides the officer an insightful view of what others see.

In the context of confidentiality and maintaining trust, many wondered whether super-visors should be given access to their subordinates' 360 information for purposes of coaching and mentoring. This scenario had critics, but there was generally less resistance to sharing 360 results with supervisors for coaching and mentoring than for evaluation purposes. Concerns included that the supervisor might not be able to set aside the information when it came time to provide annual performance evaluation comments. Others suggested that it might lead supervisors to feel as though their hands were tied because they could not take official action about negative information. People also expressed concerns that raters would water down com-ments if the information would be shared with supervisors, or people would pick only raters who would say positive things, negating the developmental benefit of honest feedback.

Even participants from the Army were also generally cautious about sharing information with supervisors, even for developmental purposes. The Army's Commander 360 is the Army's first attempt at sharing 360s with supervisors. Although the intent is that supervisors use the information to counsel and mentor only, some are concerned that it might be going too far and could shake the foundation of trust in 360 that the Army has worked so hard to develop. The Army is looking to see how participants and supervisors react to this new approach. It is cautiously optimistic that supervisors will be able to set aside the information from evaluations (as instructed) and that raters will feel comfortable sharing information candidly. But only time will tell whether that comfort with the process will be borne out.

Creating a Supportive Organizational Culture

Many 360 experts have noted that the success of 360 is partly dependent on whether raters and ratees accept and believe in the value of the process (see, for example, Bracken, Timmreck, Fleenor, and Summers, 2001). In line with this sentiment, several interviewees discussed the importance of cultural change management, which happens slowly. For example, in the Army,

> [MSAF 360] didn't get embraced rapidly, but it is largely part of the process now. It is now more accepted. Now, Command 360 will also go through that process.

> Part of it is an inculcation piece. It took [the Army] from 1997 to 2008 to implement. It's a cultural change.

Our interviewees expressed very different levels of readiness and openness to a 360 program. Some services have had very limited pilot programs ongoing and view the 360 as a niche instrument targeted at specific populations of service members. These individuals were reticent about the viability of a broad-based 360 instrument and whether such a tool would bring about positive or negative impact to the service overall:

> The comfort level and the data behind the value to the culture [are] not clear. It will lead to culture erosion problems, e.g., bad-mouthing or anointing each other as kings and queens. If it is done with 280 [general officers], it is possible—but for 70,000 officers, it is harder to communicate, despite resource concerns. It could be useful if you could convince them to use it correctly, but it is difficult.

Several interview participants discussed "a culture of feedback." Generally there is consensus that fostering a culture of personal responsibility and bystander intervention is valuable. However, some individuals were skeptical whether the 360 was necessary, sufficient, or even effective at creating such an atmosphere:

> Now it is a core of changing the culture where feedback is common and it is not uncommon to say, "Hey you need to check that language, or the attitude." There needs to be a mechanism for correcting behavior. A coach or self-assessment is not going to correct the behavior as much as someone grabbing them up by the scruff of the neck and making the point.

> I think we could improve as an officer corps by instituting a culture of feedback; even though it is uncomfortable, it could be helpful. And if done properly, we could get out of

this "How do you help bystanders intervene?" Feedback is another form of bystander intervention, on the really low side. A culture of feedback is worthwhile, but 360 is not a single answer. I think there [is] a series of tools that we could use that would be worthwhile. Not every year, not all grades because of burn out.

The services have different levels of experience with a process like the 360. The Army is furthest in the process and still believes that it has not yet fully integrated 360 into the culture. Many of the other services were worried about too much value being assigned to the 360, and the refrain "not a silver bullet" was often heard. Many participants did not feel that their services were ready for a culture shock as big as 360s force-wide.

Experts also point to cultural readiness as an important factor in the success of 360 efforts. Brutus and Derayeh (2002) identified challenges to implementing 360s, including resistance caused by a lack of trust in the system (e.g., a fear of being identified by peers or a supervisor), a lack of strategic alignment with organizational objectives, and poorly designed rating instruments. In another study, managers at a university in the United Kingdom were concerned about a lack of support from supervisors for developmental activities and a mismatch between the competencies addressed in the 360 program and those emphasized in other organizational policies, such as criteria for promotion (Mabey, 2001). The organizational environment must support feedback-seeking and feedback-giving, which may not be the norm in more-hierarchical organizations (Funderburg and Levy, 1997). All of these are examples of issues that might make large-scale implementation of 360s in the services particularly difficult.

Researchers have also shown that leadership's attitudes toward feedback processes can influence employees' reactions to the feedback (see, for example, Atwater and Brett, 2005). This provides further support for the sentiment that ensuring buy-in from leaders and supervisors in the services before implementing 360s more broadly would be critical to ensuring success. According to the comments from our interviews, that type of buy-in in the services could be developed gradually over time (as evidenced by the Army) but would take years, and forcing 360s on the services when they are not ready for them could lead to less buy-in rather than more.

Obtaining Accurate Ratings

One of the most frequently mentioned concerns, even for developmental purposes, was related to the accuracy of the information that would be obtained. Concerns included potential variability in results depending on which individuals participate in the review, how to evaluate the quality of subordinate input, the importance of contextualizing results within an individual's chain of command, and the need to vet and approve the 360 questions among individuals who will participate in the 360 review. Some of these issues have also been the focus of many empirical studies.

Our interviewees believe that honesty is a big concern, particularly when using 360s for evaluation, and experts express similar concerns (see DeNisi and Kluger, 2000; London, 2001). Participants expressed concerns that raters might whitewash their comments for a variety of reasons, including fear of reprisals, friendship with the ratee, and loyalty to the ratee. Others might attempt to game the system or respond quid pro quo. Peers might be motivated to undercut someone they see as competition, and subordinates might conspire to get a supervisor they dislike fired, particularly if their own jobs were in jeopardy. Still others might not take the time to respond carefully or accurately. It might become just another "check the box" activity.

The consensus of our interviewees is that using 360s in evaluation could change responses in so many ways that the results could ultimately be meaningless and, in some cases, harmful. People could be passed over for promotion because of false information, and others who are toxic might still slip through undetected. In the words of one of our participants,

> Conventional wisdom in regards to 360-degree assessments from experts and researchers is that the most effective use of 360 assessments is to enhance professional, individual development. Once you change the purpose or intent of a 360 from development to evaluation, you affect the willingness of raters to provide candid or unfettered feedback. Raters seem to be more willing to provide honest feedback if the ratings remain confidential and will be used only by the person rated for [his or her] development. When promotions are at stake and results may be distributed to others, raters tend to rate more positively.

The research on agreement among different sources of raters in 360 programs and whether ratings are consistent and stable within rater sources is particularly abundant. For example, a meta-analysis by Viswesvaran, Ones, and Schmidt (1996) found somewhat higher interrater reliability for supervisors than for peers. Similarly, a meta-analysis conducted by Conway and Huffcutt (1997) found that reliability was highest for supervisors and lowest for subordinates, with peers in the middle. In general, however, the reliability coefficients are low for all sources. A number of studies have shown that reliability is also influenced by characteristics of the job, with greater agreement among sources when raters have more opportunities to observe ratees or when jobs are less complex or have more concrete or observable behaviors (e.g., Harris and Schaubroeck, 1988; Heidemeier and Moser, 2009; Warr and Bourne, 1999).

Research findings on agreement between different sources of raters is mixed. Meta-analyses have generally found low correlations for self, supervisor, peer, and subordinate ratings, although correlations also varied among different pairs of sources, with higher correlations between peer and supervisor ratings than between other pairs of rater sources (Conway and Huffcutt, 1997; Harris and Schaubroeck, 1988). Conway and Huffcutt (1997) contended that low correlations should not be interpreted to mean that some sources are inaccurate or should be excluded; instead, the authors concluded that low correlations among different sources demonstrate that the different sources provide unique perspectives on performance.

Likewise, others have argued that ratings from different sources should show low to moderate agreement; high levels of agreement would be redundant and provide little benefit given the costs of collecting 360 feedback (Borman, 1997; Tornow, 1993). Other research (e.g., Facteau and Craig, 2001; Scullen, Mount, and Judge, 2003; Woehr, Sheehan, and Bennett, 2005) has shown that different types of raters have similar conceptualizations of performance, supporting the notion that meaningful comparisons can be made across sources. More recently, LeBreton and colleagues (2003) raised questions about the use of correlations to assess agreement among different rater sources in light of restriction of range that is common in performance ratings. Using an alternative approach, LeBreton and colleagues demonstrated much higher congruence among raters than previously documented.[1] Nevertheless, it is possible that

---

[1] The authors concluded that ratings from different sources may provide fewer psychometric benefits than previously thought but may provide psychosocial benefits, such as increased job satisfaction, organizational commitment, and percep-

ratings from some sources are inherently more accurate than others.[2] For example, Atkins and Wood (2002) showed that self-ratings can sometimes be the least accurate source in a 360. And for at least some aspects of performance in the military (such as leadership potential), peers might be accurate predictors of future success.[3]

Another aspect of ratings that reflects some of the interviewees' concerns about 360s is leniency. In their meta-analysis of self and supervisor ratings, Heidemeier and Moser (2009) found that leniency was moderated by the purpose of the ratings, with greater leniency in ratings conducted for administrative and developmental purposes than in ratings conducted for research.[4] Surprisingly, leniency appeared to be similar in ratings conducted for developmental and evaluation purposes. However, a variety of other studies show that raters are more lenient when ratings have a direct impact on ratee outcomes—such as their use in promotion decisions. In a study of undergraduate project teams, Farh, Cannella, and Bedeian (1991) found more leniency in self and peer ratings when ratings were used to determine grades rather than for developmental purposes. In addition, peer ratings in the developmental condition showed less inflated ratings, better discriminability, and higher reliability. In research on performance evaluation more generally, a meta-analysis showed higher ratings for administrative decisions than for research (Jawahar and Williams, 1997). Longenecker, Sims, and Gioia (1987) found that executives reported inflating performance ratings for a variety of political motives, including reasons related to employee welfare, such as job benefits (e.g., to maximize the employee's merit increase) or career opportunities (e.g., to avoid creating written documentation of poor performance). These findings support arguments favoring 360 feedback for developmental rather than evaluative purposes. When it came to our interviewees, participants wholeheartedly agreed:

> My concern is that you will eliminate the efficacy of the tool. People will no longer be frank and honest.

---

tions of organizational justice. Fleenor and colleagues (2010) noted also that these findings suggest that self-ratings should not be compared to separate sources of ratings from others but to combined ratings across other sources.

[2]   The accuracy of 360 ratings is a question of *construct validity*. Construct validity typically is established by analyzing correlations among variables that purport to measure the same concept or phenomenon. For jobs with objective measures of performance, one may be reasonably confident in the construct validity of 360 ratings. However, because "true" performance is often unknown—particularly for managerial or leadership positions for which there are typically few objective performance criteria—the construct validity of 360 ratings has been assessed by correlations with other/external measures of performance. However, it is important to bear in mind that these external measures may themselves be subject to various confounds or other factors that limit their reliability or validity.

[3]   For example, a study of U.S. Army officers found that peer ratings of potential for advancement were predictive of ratees' promotion to general (Downey, Medland, and Yates, 1976).

[4]   In addition, Heidemeier and Moser (2009) found that rating-scale format and content affected leniency, with reduced leniency when social comparison scales were used, when ratings focused on task or contextual performance rather than on traits (such as personality characteristics), and when ratings consisted of global measures rather than separate dimensions of performance. (*Task performance* focuses on meeting the technical requirement of one's job; *contextual performance* addresses behaviors contributing to the broader goals of the organization, such as helping others and following organizational procedures.) Warr and Bourne (1999) also showed that ratees gave themselves higher ratings than other sources did on more-subjective criteria (interpersonal skills and flexibility) and lower ratings than others on more-objective criteria (technical knowledge).

A concern with rating scales is that scales are already skewed. Only the very worst performers get anything but the highest score. [5]

There is probably invalid kind of stuff on both ends of the spectrum. That's why it needs some way to be seen overall and contextualized.

You have to watch the [phenomenon] where you have to get ten subordinates who all get together and all collude to rate someone badly especially if they know it is to be used for promotion. Especially subordinates who have an axe to grind. If they knew that the 360 would be used for promotion, they could use it wrongly. Probably we would say *no* to evaluation purposes [for this reason].

Harming the Organization When Using 360s for Evaluation

There are a number of ways that 360s could do more harm than good. Resources could be wasted on programs that are ineffective, or people might focus their performance improvement efforts on areas of performance that an organization does not value. However, two areas of harm are of greatest concern when used for evaluation.

Could Ruin the Value of 360 for Development

Several participants expressed concern that the trust that people currently have in 360s for development could easily crumble if the services even implied they might be considering using them for evaluation. The Army has worked for years to develop trust from personnel that the feedback provided to others will be anonymous and that the ratings about an individual will not be shared with anyone else. Interviewees both within and outside the Army are concerned that it would take many years to rebuild that trust, were it to be broken.

Participants also mentioned that asking participants to complete two separate 360s, one for evaluation and one for developmental feedback, would not be a viable solution. Doing so would be costly to the raters, but even more importantly, participants believed that raters would easily become confused and eventually they would approach both as evaluation.

This is consistent with concerns expressed by many experts in personnel research. For example, in an article on how to help improve the effectiveness of 360 feedback, DeNisi and Kluger (2000) cautioned that

> many organizations that start out to use these appraisals only for developmental purposes use them eventually for decision making, as well. In fact, we are familiar with several organizations that introduced 360-degree appraisals strictly as tools for development, but have either begun using the ratings for making some types of decisions, or they are considering doing so. To do so, especially after stating that the ratings would be used for developmental purposes only, is likely to hurt an organization's credibility with its employees, which can lead to many problems and reduce employees' willingness to change their behavior following feedback. (p. 136)

---

[5]   Note that performance evaluations in the military have long suffered from just this concern. Inflated ratings would be even more problematic with 360s because people would fear retaliation by supervisors.

Could Lead to an Unwillingness to Discipline Subordinates

The phrase "popularity contest" arose often in our interviews. For some, this was the topic that most occupied their thinking about 360 evaluations. The general concern is that 360s can reflect how much a subordinate likes his or her superior but that like or dislike is not necessarily relevant to accomplishing the mission. Some are worried that officers may concern themselves with how they are perceived and may be less likely to make unpopular choices when the mission demands it:

> You might be getting something bad too: A leader putting his personal ratings above the hard decisions he has to make to get the mission done.

> Oftentimes the commander is expected to come in and right the ship, and in doing so can fairly, properly, and appropriately make some tough calls or shift a culture on a given base. [Commanders] make things change that you might not like. So to some degree, the 360 is also a bit of a popularity contest.

> If everybody knows it is a development tool, it will lend itself to honesty and to people improving themselves. If it gets into the performance evaluations, I hope it also doesn't lead to people trying to do something just because it gets good scores.

Additionally, some mentioned that 360s could create a perception (real or not) that disciplinary actions are taken as reprisal for negative 360 ratings. Others felt that they could lead to excessive investigations and ultimately a distrust and fear of the process:

> There is no recourse for a [general officer], so to me, what we are doing is making it difficult for leaders to discipline soldiers for fear of being investigated. What they tell you is that, if you haven't already been investigated, it will happen soon. It causes a lot of grief for families and leaders when people come at them again and again.

All of these are concerns are among those raised by experts as well (see London, 2001).

Although many of our participants raised this as a big concern, a few also followed up by acknowledging an entirely opposing view. They explained that truly good leaders might still do the right thing and make the tough calls in spite of the potential for their actions to lead to low 360 ratings:

> Most commanders don't want to be disliked but want to make the right decisions because happy troops are productive troops. If we have guys [who] don't make tough choices, then we didn't pick the right commander to start with. I don't think most people will be seduced by this. I think most people will continue to do what they think is right and the hell with what happens.

Most of those who offered this alternative view were expressing a clear uncertainty about the consequences of 360s. One participant, however, did not see it as an uncertainty at all. They instead suggested that, although this is often presented as an argument for not implementing 360s, leaders in the military simply would put the mission first.

Making 360s Logistically Feasible and Conserving Resources

Our interview participants were concerned about the feasibility of a far-reaching 360 system. The concerns focused largely on the difficulty of managing such a large system, the challenges of navigating policy to support such a program, the costs of implementation, and the survey fatigue that service members may feel as a result of it.

Representatives from some services mentioned they lack the capacity to scale up quickly. We heard concerns about data management, data security, storage, and responsibility associated with building such a large system. Some pointed to the experience of the Army and the rethinking that was required to sustain its 360 program as it attempted to scale the system up. Additionally, there were concerns that a service-wide 360 system would require greater cross-service interoperability than is currently available. Having raters outside of the ratee's service would likely be common, but the current information technology systems do not connect well across services. The Joint 360 is currently grappling with exactly this challenge. In addition, some noted that implementing this for evaluation would require significant alterations to the existing technological systems for evaluation to accept this sort of new input. Such a change to the evaluation systems would necessarily have to happen slowly.

Coaching was another logistics and cost concern. Finding, training, and retaining the required number of coaches needed to take full advantage of 360s would be costly. Participants questioned whether it would be logistically feasible or cost-effective to provide coaching to all officers (see the next section). If 360s were administered on a large scale, many felt, supervisors would be the only viable option, but they would need to be trained before they could take on such a role; however, few thought that sharing the information with even trained supervisors was a good idea (see the earlier section on coaching for more discussion on this). The Army's solution with the MSAF is to offer coaching to any who want it. It has found, however, that only a small number of participants take up that offer, thus keeping costs down. Given that research has shown that the utility of the 360 could be significantly undermined without some sort of structured follow-up action, some have asked how useful it is to administer it to the large majority who do not meet with coaches.

Although coaching and administration costs alone might be a deterrent for administering 360s force-wide, raters' time costs were an even bigger concern. Our participants estimated that it takes anywhere from ten to 30 minutes to complete the various 360 rating forms they are currently using, but several also noted that it usually takes significantly longer for those who add written comments or for those respondents who take the time to do it well. Multiply that by the number of people above and below each service member, and the number of man-hours spent filling out forms about other service members would be significant:

> We are always conscious of making sure that what we do is worth the [service member's] time to do it. We tried to minimize some of our total force awareness training to get that into a smaller window. We need to be mindful of whatever we assign takes up time.

> I was a commander and I had seven squadron commanders underneath me, so I [had] to do those seven because I'm their supervisor. There are ten other colonels on base, so I've got to do them because I am their peer. And then my bosses, I am their subordinate. And now you are looking down going, "we aren't going to get any work done."

> If I am a commander, I am going to be asked to fill out surveys on every one of my officers because I am their supervisor. Or I am group commander or wing commander and it cascades. I mean, we could be in the survey business. That is one that I really struggle with, is the sheer volume.

Several also mentioned that, given the current work obligations of most officers, 360s would likely have to be completed in their off hours at home.

Although some efficiency could be achieved by not having everyone complete a 360 on everyone else, it would still likely be burdensome. As the Army has discovered, when raters are chosen by the ratee, some people are picked as raters much more often than others. As a result, the numbers of ratings those people have to complete about their peers, supervisors, and subordinates can be quite large. In addition, researchers have shown that a minimum number (at least six or seven subordinates and four to seven peers) is needed to achieve good reliability (Conway and Huffcutt, 1997), and, to help protect anonymity, more is better.

Oversurveying and survey fatigue were frequent concerns, particularly because of the variance in unit structures within and across the services. In some units, an individual is in command of six officers; in others, an individual is in command of over 100 officers. The obligation to fill out 360s on that many individuals would be overwhelming:

> I haven't had enough input from raters to complete my survey. . . . A lot of the individuals really don't have the time to do it.

> We get surveyed to death, to the point where people try to avoid them. You only fill it out when they've come to you six times. This will just be more of that, and it is really detrimental because when we really need information we can't get it because people are just tired.

> Just thinking about the time it takes to fill out an MSAF . . . just multiply that. I don't know if that's a fair expectation of young soldiers, airman, sailors, and Marines. To [take that amount of time] away from their training, away from their duty day.

The Army is already seeing the impact of that survey fatigue, with many participants unable to obtain the minimum number of ratings because the raters are unwilling or unable to take the time to complete them.[6]

Additionally, there was concern about the frequency of 360s. If the program were linked to evaluation, then the 360 feedback would have to be gathered and processed annually before a hard deadline. At present, the Army expects officers to complete an MSAF only once every three years. As a result, expecting it to be completed any more frequently than that would likely be seen as highly burdensome, even in the Army.

> That's another piece—frequency. How often do we do this? If you are new in your job, do I wait a year before I give it to you? And then do I give it to you every other year? Every three years? These are policy questions that will drive cost and implementation.

Similar concerns about time requirements occur in civilian-sector applications. Brutus and Derayeh (2002), for example, found that the time and resources required to implement

---

[6]   See the earlier discussion on this in Chapter Four.

360-feedback programs was a major deterrent. Likewise, in post-360 surveys, Edwards, Ewen, and Vendantam (2001) found low agreement that 360s were an efficient use of time.

A few noted that ultimately it was a question of prioritization of human resources over hardware, and the trade-off might be worth it:

> For [the cost of] one attack helicopter, the Army can [pay for 360s] for ten years. We are way underresourced for selecting, promoting, and taking care of our people. I think the cost, if the program is properly put together, is justifiable.

Others felt that it would just come at too high of a cost. Some other aspect of people's time would have to be traded.

They also expressed concern that the expense along with workload demands would lead to cutting corners. A need to make the rater forms short, which would eliminate much of the valuable detail; the fact that people would be overburdened and would stop putting thought into the ratings and comments; and a lack of meaningful interpretation, follow-up, and developmental assistance were some of cost-cutting measures they foresaw—all of which would negate the value of administering it in the first place:

> I just think logistically . . . once we start to try to implement it, it would almost be infeasible. So then you would force us to go "we have to do this." So then what you would have to do is cut corners.

> If you want it quick, you are going to get it quick. And if you get it quick, then are we filling the square that we did a 360 and missing the true value.

> How do you train people to use the data professionally? Or would you just look at the data and go "well, it looks good to me; no one really complained about you"? Versus the trained coach [who] sits down [and spends a lot of time going over it with the ratee].

Including Better Measures of Success

Measures of the success of 360s in the services are fairly limited. Many of our participants had a number of positive things to say about how useful they thought their own experiences receiving 360 feedback had been. Some described eye-opening experiences in which the detailed comments brought to their attention certain aspects of their behavior they had never previously realized might be problematic. A few, on the other hand, felt that the results were not particularly helpful or enlightening because they did not learn anything new that they did not already know. However, reactions to 360 feedback can be positive, even when there is no measurable improvement in performance.

Perhaps the most important question regarding the use of 360-degree feedback is whether individual and organizational performance improves as a result of this technique and under what circumstances. So do 360s in the military lead to performance improvements? That question remains largely unanswered. Discussions with DoD stakeholders suggest that most 360 efforts currently in place do not have any existing method for evaluating this. Some (such as the MSAF) collect satisfaction ratings from the person who was rated as their primary outcome measure. Others, however, have no outcome measures in place at all. Many agreed, however, that such measures would be useful:

> Okay, a year later, six months later, you worked up a performance plan with a coach. Did you ever implement that? Do you have examples in the past year as a supervisor, leader, commander that you practiced what was identified in your 360? I don't have [those] data, but [they] would be interesting.

One of the problems in determining whether individual performance improves is that it would require a systematic data-collection effort. And taking the time to do that is just one more burden to add to the 360 process. Nevertheless, when we asked whether the services had considered how to measure success, many agreed that such measures are needed and are important in evaluating whether 360s are worth the cost:

> Why 360s? How will they make us better [individually, collectively]? Supporting evidence inside or outside DoD? There is an assumption in some circles that it can be used to root out toxic leaders—what is the effect on the climate? . . . For development, what return on investment is the Army receiving?

In one discussion, the idea was raised of taking 360s at regular intervals during the course of a career. This longitudinal approach would provide service members an opportunity to see if they have improved in certain areas over time. For example,

> if every three years I take a 360, and I can see that I consistently take a hit in whatever. Maybe I am consistently very fair but horrible communicator. That's hugely valuable. I would rather have [those] data than water [them] down with ratings. That is a huge opportunity, but logistically that is challenging.

However, in discussing whether 360s could be used to compare someone's ratings over time to see whether there was improvement, some participants agreed that it might be useful but noted that such a longitudinal look was not currently possible:

> The reports do not provide a comparison tool [for results over time]. A longitudinal look would be very valuable. Also, because it's so recent, it's going to take few years before someone can see the change.

But even this longitudinal approach has shortcomings because only the individual can see the results, so it is difficult at an organizational level to know whether individuals are improving as a result of the use of 360s or not—and to, in turn, determine the benefit to the organization. In addition, other participants pointed out that people change positions so frequently in the military that differences from one 360 administration to the next might just simply reflect a change in the people providing the ratings or new expectations in a completely new job. As a result, a longitudinal look might prove inconclusive.

Finally, one interviewee suggested using unit 360s or command climate surveys as a tool for measuring behavioral improvement.

> I think that is best handled by a climate survey. . . . Most senior commanders get briefings periodically on numbers of people, money left in the bank, how many tanks are running,

how many aircraft are down for various things. But only irregularly do we get the perception of organizational readiness of a unit. . . . If I were king, I would say start out either simultaneously with or even ahead of this individual 360 and do a 360 on your discernible organizations, my hope is that it would be done at the same time. I think the best way to measure the collective impact of leader development efforts is by looking at how they play out in the real world, and the real world can be assessed rather simply and not very expensively by periodic climate surveys.

Tailoring 360s to the Services' Needs

The need for service-level autonomy over implementation decisions was mentioned during several interviews and from respondents from all organizations:

> [Human resource] systems need to be tailored to each of the services. The services have different cultures and they are different for a reason. It may match one but be a [forced] fit for others. The Army may want someone who follows commands, while the Air Force may want a technical specialist who thinks outside the box and can think of a number of ways to break into a database.

Participants also noted that, even within a service, a one-size-fits-all approach will not work well. Who participates as raters and whether it makes sense for some people to even receive 360s are important questions that should not be overlooked:

> I think you need to take into account some career field differences. You won't be able to ask about leadership for some of the occupations and younger folks.

> When you have [an] Army or Marine [Corps] construct, where the organization is a lot flatter, and you got platoons, and companies, and battalions, there are more leadership opportunities, more opportunities to manage people.

They also discussed a need for flexibility in other key design, administration, and technological features, such as the opportunity to customize questions to occupations and grades:

> For it to be effective, the officer has to be in a leadership environment. I don't know how you can ask the same kinds of questions. Maybe you could have a tailored piece for different ranks. . . . Even in operational areas, a young fighter pilot trying to get upgraded to instructor pilot, he is flying, he is supposed to be in mission study, he is supposed to be flying as much as he can; how do you assess his leadership? He is a little different than a section commander in a thousand-person maintenance squadron.

In addition, the services need to have enough flexibility to decide who was in a position to benefit most from a 360. There was consensus among the majority of our participants that 360s would be most useful and much more cost-effective if they were implemented only for personnel in higher-level grades and positions:

> I would think you want to hold off until field grade O-4–O-5. This is really because of the time involved. The overhead is too high time-wise.

Development should start much lower level; feedback for selection should start at O-6 and O-7.

I think 360 for development is a wonderful thing, but we can't do it for everybody and should focus on the midgrade officer and higher. Captain is good place to start.

Money, time, fatigue. Personally I don't think it's worthwhile. All ranks need development, just like a general officer. . . . [But] the feedback [gotten] from the population (particularly subordinate and peers) won't be as helpful for junior people. The senior look is helpful for anyone.

From O-2 or O-3, we have the problem of an unstable subordinate population—they move around too fast—and the relationship could be too close here to be done effectively without serious professional guidance.

Finally, some of our participants commented that, while officers are certainly an important group to include in 360 efforts, they are not the only relevant group. It would be useful to consider including enlisted personnel and civilians, but when to incorporate them and how to do so should again be left up to the services' discretion.

CHAPTER SEVEN

Implications for the Military Services

When 360 programs are implemented, they should have a specific purpose in mind that aligns with organizational goals. Without a well-understood purpose, it is not possible to know whether the system is working and is worth the time and resources that an organization must invest. This applies to private-sector entities as well as the military. So, the first step in introducing or expanding use of 360s is to determine its purpose.

Currently, 360s in the military services are used for developmental purposes only, not as an evaluation of past performance. The overall consensus from our interviews is that the primary purpose of 360s should continue to be for development, particularly for officers in leadership positions. In this context, results are intended to help an officer hone his or her skills as an effective leader by giving an individual the opportunity to develop goals and courses of action. In this chapter, we conclude with our assessment of whether 360s are the right tool for the U.S. military.

Purpose and Design Matter

According to Bracken, Timmreck, Fleenor, and Summers (2001),

> Multi-source feedback looks deceptively easy. It is like watching a proficient athlete and saying "I could do that" without any knowledge of the training and preparation required for excellence. In much the same way, an effective 360 feedback system that creates sustainable, focused behavior change requires both good tools and good implementation. Systems that fall short in one or more of the [key features] will probably not "hurt" anyone, and may actually survive in a benign way. The problem is that they will not create behavior change, at least in the manner envisioned by the organization. At best they are a waste of money. At worst they divert time and resources toward activities that will not bear fruit and may cause employee morale problems.

Perhaps the most important step in adopting or expanding the use of 360 feedback—in the military or elsewhere—is to determine and provide guidance on the purpose of the program. In the military, for example, these assessments have been used primarily for force development—to help build better leaders and help individuals understand gaps in how they perform their jobs. Congress has asked whether there is a role for 360 feedback in the appraisal system. If used for evaluation, what would be the purpose? Some interviewees speculated that Congress's intent for including 360 in evaluations is "to find the toxic leaders or the toxic leadership characteristics that we might not know otherwise."

But as we heard during our interviews, Congress's reasoning for implementing 360 for all officers in the military services has neither been clearly expressed nor is it well-understood: "What are we trying to fix here?" "I've yet to see someone who comes and, through a military lens, says, 'we are going to do 360s, here are our objectives, here is our plan.'" If the services were to implement 360s force-wide, they would need a statement of goals that clearly articulates what the program is meant to accomplish—something more specific than "get better leaders" or "screen toxics."

Specificity of goals is important because views diverge on the utility of 360, as the discussion in the previous chapters illustrates. Some believe that existing tools can accomplish broad goals, such as "better leadership," in a more convenient and cost-effective way:

> I think there is a convenient and powerful role for organizational climate surveys that take care of a lot of this, and [they are] so much easier to do in the short term. How do you measure progress? It is going to take a while, but, by conducting climate surveys over time, you see trends, and instantly you can get a good feel for what is good and bad. Climate surveys are much less threatening to a commander than a subordinate assessment. Climate assessment can provide a lot of useful information for discussion for how things are going in the organization that can have a side impact of quality of leadership and so forth.

Others feel that 360s can make a contribution as part of a suite of tools:

> But we also think that there is a way to use 360, command climate surveys, and other tools to understand the impact of the commander on [his or her] unit. The 360 is one tool to use to understand how leaders are doing in command. But we also think [that] these tools are useful to develop people before they get to command.

Regardless of how the services decide to use 360s—and we anticipate that the goals and use could and probably should differ by service—a clear statement of goals is an essential starting point. Without such clear statements, it will be impossible to evaluate whether the program is designed properly, working as intended, or even needed at all.

Recommendations for Using 360s in the Military

"No" for Evaluation

Based on our research on 360s, both within and outside a military setting, we advise against incorporating 360s in the officer evaluation system at this time. Although the results of our interviews suggest that implementing 360s force-wide for evaluation purposes would *technically* be feasible—that is, systems could be changed, policies could be revised, new guidance could be issued, and personnel could be trained to handle the use of 360 feedback information in military performance evaluations—it simply would not be advisable for several reasons.

For one, it could ruin developmental applications, and the Army's experience shows how long it can take to develop trust in such a process. Trying to implement two separate systems could be confusing to raters and increase survey burden on the force. Using 360s as part of evaluation processes could also create perverse incentives for commanders who, thinking about the impact of 360 feedback on their performance appraisals, could make decisions (consciously or unconsciously) based on a desire to be liked by the troops that might be inconsistent with the mission. There are similar concerns about incentives for raters who could sabotage reviews in an attempt to get rid of an unpopular leader or could whitewash ratings for friends or for fear of retaliation.

Moreover, there are a host of logistical issues that are unresolved, perhaps the most important of which is the need to develop formal guidance in the use of the tool. Because of the complexity of using 360s in the evaluation system, mandating their use could lead to largely useless or meaningless ratings that are difficult to interpret without considerable context. The performance evaluation system fundamentally provides a record of past performance. It also signals how well the officer's performance aligns with the traits that are valued by the service. Should 360s become part of the formal evaluation system, clear guidance on how to treat the results would be required when there are differences among superior, peer, and subordinate assessments.

Perhaps the key concern in using 360s beyond developmental purposes is the potential impact on selection boards and the promotion process. For the purposes of determining the effect of expanding performance evaluation reports to include a 360-degree assessment, the provision of law that would directly link the assessment to selection boards and the promotion process is Section 615 of Title 10 of the United States Code. While a requirement to include 360s in performance evaluations may not reference using that information for promotion consideration purposes, an officer's evaluation reports are one of the primary documents reviewed by board members. Thus if the results of 360s become a required element of evaluation reports, that information would be provided to selection boards, absent a change in service policy as to the final disposition of performance reports, and would be used during the board's deliberations.

The inclusion of this information in board deliberations is of concern for several reasons. The information from raters is usually anonymous and therefore cannot be challenged by the ratee. Information in 360s can sometimes be inaccurate because of rater lack of care or skill in providing ratings. In a high-stakes situation, such as promotions, raters could be dishonest in attempts to positively or negatively impact board selection decisions with no potential for recourse. Finally, 360 ratings often require an understanding of the context in which the ratings were provided. A difficult assignment might lead to low ratings despite exceptional performance on the part of the ratee. Such context would not be obvious to the board and has the potential to worsen rather than improve the accuracy and fairness of promotion decisions. Such information should be included in board deliberations only after careful examination of its accuracy, both initially and over time.

We acknowledge that there may still be a desire for new avenues by which to incorporate input from peers and subordinates into the performance evaluation system. Though we did not explore specific options in this study, other approaches to doing so could be worth considering. For example, the evaluation rating form itself could be redesigned so that raters are required to comment on what peers and subordinates think about the person being rated—formalizing the informal inputs that many supervisors already gather from peers and subordinates. Encouraging this type of information-sharing and training raters on how to solicit and use such information is one way for the performance evaluation system to capture a broader range of perspectives. Although this type of alternative may be worthwhile, we cannot make a specific recommendation because the aim of this study was to determine whether including 360s in performance evaluations would be advisable. We do not believe that 360 assessments are the appropriate tool for this purpose—at least not at this time.

"Yes" for Development, but Proceed with Care

Based on the literature and the interview findings, we would advise the use of 360s for development purposes. 360s could be used for leader development and made available as a service to individuals hoping to improve. The tool could be offered to anyone who wants to use it, along with coaching to help service members evaluate the results and incorporate them into self-improvement goals. 360s could also be adapted for special uses, such as in training courses.

In addition, 360s could be used to provide an aggregate view of leadership across the force—something that current tools are not well positioned to provide. Individuals could compare their performance with that of other service members. Leaders could use the 360 to identify groups (such as units, occupations, or bases) in which ratings are consistently better or worse than others (regardless of the person in charge). The feedback could be used to identify areas that could benefit from developmental assistance or other support interventions. The results could also be used to identify force-wide strengths and weaknesses in performance. For these purposes, the tool could be implemented for a sample of the force, not as a census, if survey fatigue is a concern.

Such an aggregate use of 360 information was mentioned briefly by only a few of our participants. Nevertheless, this application is in keeping with the well-supported goal of using 360 feedback for developmental purposes. To the extent that 360s could be used to supplement or complement existing or future climate assessments, the idea has merit. Yet care should be taken to ensure that any overlap with existing climate assessments is intentional. Otherwise, the effort could be duplicative of efforts already under way and hence a waste of raters' time.

The bottom line is that making 360 feedback available for developmental use in the military services is a good idea and is essentially how the tool is being used today. Each of the services uses 360s in ways that it finds useful in support of its personnel management goals and in line with service culture. While use of 360s could be expanded in the services, how and when that is done should be determined by the services themselves. On the other hand, OSD could consider using a similar tool to solicit input at an aggregate level to survey personnel leadership skills across the services, similar to the use of climate surveys today.

A mandate to use 360s department-wide could waste significant resources. A one-size-fits-all approach would be ineffective in most cases because the needs of the services, of individual service members, of particular career fields, and even individual positions differ. Moreover, other tools are in place within the services that might be achieving similar goals. So further implementation needs to consider efforts already under way and what 360s would add. And although cost of the tool itself is relatively low, time costs for personnel (particularly to get maximum benefit of 360s) is high, so the purpose of implementing the program and how it aligns with organizational goals need to be made clear.[1]

Our assessment suggests that many features of the Army MSAF program (such as service-specific competencies and rater qualifications) would likely need to be tailored for implementation in other services, even though it is the most developed among the service programs and a useful model for how 360s could be used. Although the Center for Army Leadership

---

[1]   Participants with experience administering 360s in the services cited times to complete forms as ranging from ten to 30 minutes. However, they noted that, for those who provide comments, the forms can take much longer to complete. Time costs for coaching and follow-up after receiving the results were not estimated, but those who had experience receiving intensive coaching in a developmental setting spread over days or weeks described that scenario as the ideal but time-consuming.

has offered to adapt the MSAF and provide additional support to address any service-specific needs, the services also might want to develop their own tools and in-house support staff.

Nevertheless, as shown by the Army's existing MSAF effort, providing the needed infrastructure to support the tool is no small undertaking. An experienced staff is needed to develop and maintain the survey content, supporting materials, and technological interface; review survey responses and requests for data; provide coaching and follow-up; evaluate success of the tool; and more. Some of the services already have existing small-scale efforts that are supported by either external vendors or in-house staff. Those efforts could be expanded, although changes would be needed to adapt them to other populations or uses, requiring larger in-house staff or further negotiation with vendors. Regardless, the Army's existing tools are available to anyone with a CAC who wants to use them. This means that anyone in the military can voluntarily log in and initiate a 360 on himself or herself. As a result, the Army's tool could be adopted by anyone in the military interested in using it off the shelf.

Current Implementation Is on the Right Track

Overall, our interviews showed that the spirit of 360s clearly resonates with the services. The services clearly value good leadership behaviors and tools that can help develop good leaders, and the 360 is one tool that has value in developing leaders. Interviewees agreed that achieving the mission at the expense of people and values is not acceptable. The services want to eradicate problematic leadership behaviors, such as toxic leadership (poor treatment of subordinates) and unethical behavior (fraternizing, sexual harassment, and assault). But there is a general consensus that the 360 is not the right tool for these specific purposes—other tools exist that can be used more effectively with fewer burdens on the force.

Stakeholders generally find 360 feedback beneficial but recognize that 360s are not a silver bullet. Experts in the field of personnel research agree and caution that overzealous use of 360s without careful attention to the content, design, and delivery can even be harmful to an organization and not worth the time and money spent on them. For example, according to DeNisi and Kluger (2000),

> the design of any feedback intervention requires thought and consideration of the factors that are likely to make it more effective. Without such consideration, organizations might actually be designing systems that are not cost-effective in the long run. (p. 138)

Each of the services has used 360 assessments to some degree—with the Army having the most experience to date. The services have commissioned many reviews on this topic, and many service members have chosen 360 feedback as their thesis topics at the various command and staff colleges.[2] Nevertheless, while they agree that the tool can be effective in some circumstances, the services also see other ways to solicit input from subordinates on leadership behavior. Moreover, they believe that mandating 360s force-wide, even for development purposes, is not necessarily the right answer to solving leadership problems within the services. Service cultures may not be equally supportive of 360s, so mandating their use could lead to failure. Rather, it is more advisable to allow the services to continue on their current paths, expanding the use of 360s in a way that is tailored to individual service needs and goals.

---

[2] For some examples, see Fiscus, 2011; Fitzgerald, 1999; Hancock, 1999; Leboeuf, 1997; Nystrom, 2001; Pointer, 1996; Psotka, Legree, and Gray, 2007; Reese, 2002; and Williams, 2005.

Abbreviations

360	360-degree assessment
CAC	common access card
CCL	Center for Creative Leadership
DoD	Department of Defense
GS	General Schedule pay scale
MSAF	Multi-Source Assessment and Feedback
NDAA	National Defense Authorization Act
OSD	Office of the Secretary of Defense
OUSD/P&R	Office of the Under Secretary of Defense for Personnel and Readiness

References

3D Group, *Current Practices in 360-Degree Feedback: A Benchmark Study of North America Companies*, Emeryville, Calif., 2013.

Air Force Instruction 36-2406, *Officer and Enlisted Evaluation Systems*, Washington, D.C.: Department of the Air Force, April 15, 2005.

Air Force Instruction 36-2501, *Officer Promotion and Selective Continuation*, Washington, D.C.: Department of the Air Force, March 14, 2014.

Army Regulation 350-1, *Army Training and Leader Development*, Washington, D.C.: Headquarters, Department of the Army, August 19, 2014.

Army Regulation 600-8-29, *Officer Promotions*, Washington, D.C.: Headquarters, Department of the Army, February 23, 2005.

Army Regulation 623-3, *Evaluation Reporting System*, Washington, D.C.: Headquarters, Department of the Army, March 31, 2014.

Army Doctrine Reference Publication 6-22, *Army Leadership*, Washington, D.C.: Headquarters, Department of the Army, September 10, 2012.

Atkins, Paul W. B., and Robert E. Wood, "Self- Versus Others' Ratings as Predictions of Assessment Center Ratings: Validation Evidence for 360-Degree Feedback Programs," *Personnel Psychology*, Vol. 55, 2002, pp. 871–904.

Atwater, Leanne E., and Joan F. Brett, "Antecedents and Consequences of Reactions to Developmental 360 Feedback," *Journal of Vocational Behavior*, Vol. 66, No. 3, 2005, pp. 532–548.

Atwater, Leanne E., Joan F. Brett, and Atira Cherise Charles, "Multisource Feedback: Lessons Learned and Implications for Practice," *Human Resource Management*, Vol. 46, No. 2, 2007, pp. 285–307.

Bettenhausen, Kenneth L., and Donald B. Fedor, "Peer and Upward Appraisals: A Comparison of Their Benefits and Problems," *Group and Organization Management*, Vol. 22, No. 2, 1997, pp. 236–263.

Bono, Joyce E., and Amy E. Colbert, "Understanding Responses to Multi-Source Feedback: The Role of Core Self-Evaluations," *Personnel Psychology*, Vol. 58, No. 1, 2005, pp. 171–203.

Borman, Walter C., "360 Ratings: An Analysis of Assumptions and a Research Agenda for Evaluating Their Validity," *Human Resource Management Review*, Vol. 7, No. 3, 1997, pp. 299–315.

Bracken, David W., and Dale S. Rose, "When Does 360-Degree Feedback Create Behavior Change? And How Would We Know It When It Does?" *Journal of Business and Psychology*, Vol. 26, No. 2, 2011, pp. 183–192.

Bracken, David W., Carol W. Timmreck, and Allan H. Church, eds., *The Handbook of Multisource Feedback*, San Francisco: Jossey-Bass, 2001.

Bracken, David W., Carol W. Timmreck, John W. Fleenor, and Lynn Summers, "360 Feedback from Another Angle," *Human Resource Management*, Vol. 40, No. 1, 2001, pp. 3–20.

Brett, Joan F., and Leanne E. Atwater, "360 Feedback: Accuracy, Reactions, and Perceptions of Usefulness," *Journal of Applied Psychology*, Vol. 86, No. 5, 2001, pp. 930–942.

Brutus, Stephane, and Mehrdad Derayeh, "Multisource Assessment Programs in Organizations: An Insider's Perspective," *Human Resource Development Quarterly*, Vol. 13, No. 2, 2002, pp. 187–202.

Chief of Naval Personnel, *Navy Performance Evaluation System*, Millington, Tenn.: Department of the Navy, Bureau of Naval Personnel, Bureau of Navy Personnel Instruction 1610.10C, April 20, 2011.

Commandant of the Marine Corps, *Performance Evaluation System*, Washington, D.C.: Headquarters U.S. Marine Corps, Department of the Navy, Marine Corps Order P1610.7F, November 19, 2010.

Conway, James M., and Allen I. Huffcutt, "Psychometric Properties of Multisource Performance Ratings: A Meta-Analysis of Subordinate, Supervisor, Peer, and Self-Ratings," *Human Performance*, Vol. 10, No. 4, 1997, pp. 331–360.

Department of Defense Instruction 1320.13, *Commissioned Officer Promotion Reports*, Under Secretary of Defense for Personnel and Readiness, October 30, 2014.

Department of Defense Instruction 1320.14, *Commissioned Officer Promotion Program Procedures*, Under Secretary of Defense for Personnel and Readiness, December 11, 2013.

Department of Defense Instruction 5000.66, *Operation of the Defense Acquisition, Technology, and Logistics Workforce Education, Training, and Career Development Program*, Under Secretary of Defense for Acquisition, Technology, and Logitics, December 21, 2005.

DeNisi, Angelo S., and Avraham N. Kluger, "Feedback Effectiveness: Can 360-Degree Appraisals Be Improved?" *Academy of Management Executive*, Vol. 14, No. 1, 2000, pp. 129–139.

DoD Instruction—*See* Department of Defense Instruction.

Downey, R. G., F. F. Medland, and L. G. Yates, "Evaluation of a Peer Rating System for Predicting Subsequent Promotion of Senior Military Officers," *Journal of Applied Psychology*, Vol. 61, No. 2, 1976.

Edwards, Mark R., Ann J. Ewen, and Kiran Vendantam, "How Do Users React to Multisource Feedback?" in Bracken, Timmreck, and Church, 2001, pp. 239–255.

Facteau, Jeffrey D., and S. Bartholomew Craig, "Are Performance Appraisal Ratings from Different Rating Sources Comparable?" *Journal of Applied Psychology*, Vol. 86, No. 2, 2001.

Farh, Jng-Lih, Albert A. Cannella, and Arthur G. Bedeian, "The Impact of Purpose on Rating Quality and User Acceptance," *Group and Organization Management*, Vol. 16, No. 4, 1991, pp. 367–386.

Fedor, Donald B., Kenneth L. Bettenhausen, and Walter Davis, "Peer Reviews: Employees' Dual Roles as Raters and Recipients," *Group and Organization Management*, Vol. 24, 1999, pp. 92–120.

Fiscus, James M., *360 Degree Feedback Best Practices and the Army's MSAF Program*, Carlisle Barracks, Pa.: U.S. Army War College, 2011.

Fitzgerald, Michael D., *360-Degree Feedback: The Time Is Now*, Carlisle Barracks, Pa.: U.S. Army War College, 1999.

Fleenor, John W., James W. Smither, Leanne E. Atwater, Phillip W. Braddy, and Rachel E. Sturm, "Self–Other Rating Agreement in Leadership: A Review," *Leadership Quarterly*, Vol. 21, No. 6, 2010, pp. 1005–1034.

Foster, Craig A., and Melanie R. F. Law, "How Many Perspectives Provide a Compass? Differentiating 360-Degree and Multi-Source Feedback," *International Journal of Selection and Assessment*, Vol. 14, No. 3, 2006, pp. 288–291.

Funderburg, Shelly Albright, and Paul E. Levy, "The Influence of Individual and Contextual Variables on 360-Degree Feedback System Attitudes," *Group and Organization Management*, Vol. 22, No. 2, 1997, pp. 210–235.

Hadley, Stephen J., and William J. Perry, *The QDR in Perspective: Meeting America's National Security Needs in the 21st Century*, Washington, D.C.: United States Institute of Peace, 2010.

Hancock, Thomas S., *360-Degree Feedback: Key to Translating Air Force Core Values into Behavioral Change*, Maxwell Air Force Base, Ala.: Air University, AU/AWC/116/1999-04, 1999.

Harris, Michael M., and John Schaubroeck, "A Meta-Analysis of Self–Supervisor, Self–Peer, and Peer–Supervisor Ratings," *Personnel Psychology*, Vol. 41, No. 1, 1988, pp. 43–62.

Hazucha, Joy Fisher, Sarah A. Hezlett, and Robert J. Schneider, "The Impact of 360-Degree Feedback on Management Skills Development," *Human Resource Management*, Vol. 32, No. 2–3, 1993, pp. 325–351.

Hedge, Jerry W., Walter C. Borman, and Scott A. Birkeland, "History and Development of Multisource Feedback as a Methodology," in Bracken, Timmreck, and Church, 2001, pp. 15–32.

Heidemeier, Heike, and Klaus Moser, "Self–Other Agreement in Job Performance Ratings: A Meta-Analytic Test of a Process Model," *Journal of Applied Psychology*, Vol. 94, No. 2, 2009.

Heslin, Peter A., and Gary P. Latham, "The Effect of Upward Feedback on Managerial Behavior," *Applied Psychology*, Vol. 53, No. 1, 2004, pp. 23–37.

Jawahar, I. M., "The Mediating Role of Appraisal Feedback Reactions on the Relationship Between Rater Feedback-Related Behaviors and Ratee Performance," *Group and Organization Management*, Vol. 35, No. 4, 2010, pp. 494–526.

Jawahar, I. M., and Charles R. Williams, "Where All the Children Are Above Average: The Performance Appraisal Purpose Effect," *Personnel Psychology*, Vol. 50, No. 4, 1997, pp. 905–925.

Judge, Timothy A., and Gerald R. Ferris, "Social Context of Performance Evaluation Decisions," *Academy of Management Journal*, Vol. 36, No. 1, 1993, pp. 80–105.

Kluger, Avraham N., and Angelo DeNisi, "The Effects of Feedback Interventions on Performance: A Historical Review, a Meta-Analysis, and a Preliminary Feedback Intervention Theory," *Psychological Bulletin*, Vol. 119, No. 2, 1996.

Landy, Frank J., and James L. Farr, "Performance Rating," *Psychological Bulletin*, Vol. 87, No. 1, 1980, pp. 72–107.

Leboeuf, Joseph N., *Feedback: A Critical Leadership Resource*, Carlisle Barracks, Pa.: U.S. Army War College, 1997.

LeBreton, James M., Jennifer R. D. Burgess, Robert B. Kaiser, E. Kate Atchley, and Lawrence R. James, "The Restriction of Variance Hypothesis and Interrater Reliability and Agreement: Are Ratings from Multiple Sources Really Dissimilar?" *Organizational Research Methods*, Vol. 6, No. 1, 2003, pp. 80–128.

Lepsinger, Richard, and Anntoinette D. Lucia, "Performance Management and Decision Making," in Bracken, Timmreck, and Church, 2001, pp. 318–334.

London, Manuel, "The Great Debate: Should Multisource Feedback Be Used for Administration or Development Only?" in Bracken, Timmreck, and Church, 2001, pp. 368–385.

———, *Job Feedback: Giving, Seeking, and Using Feedback for Performance Improvement*, Mahwah, N.J.: Laurence Erlbaum Associates Inc., 2003.

Longenecker, Clinton O., Henry P. Sims, Jr., and Dennis A. Gioia, "Behind the Mask: The Politics of Employee Appraisal," *Academy of Management Executive*, Vol. 1, No. 3, 1987, pp. 183–193.

Luthans, Fred, and Suzanne J. Peterson, "360-Degree Feedback with Systematic Coaching: Empirical Analysis Suggests a Winning Combination," *Human Resource Management*, Vol. 42, No. 3, 2003, pp. 243–256.

Mabey, Christopher, "Closing the Circle: Participant Views of a 360 Degree Feedback Programme," *Human Resource Management Journal*, Vol. 11, No. 1, 2001, pp. 41–53.

Mount, Michael K., Timothy A. Judge, Steven E. Scullen, Marcia R. Sytsma, and Sarah A. Hezlett, "Trait, Rater and Level Effects in 360 Degree Performance Ratings," *Personnel Psychology*, Vol. 51, No. 3, 1998, pp. 557–576.

Murphy, Kevin R., "Explaining the Weak Relationship Between Job Performance and Ratings of Job Performance," *Industrial and Organizational Psychology*, Vol. 1, No. 2, 2008, pp. 148–160.

Navy Personnel Command, "Command Climate Assessment," web page, 2014. As of January 18, 2015: http://www.public.navy.mil/bupers-npc/support/21st_Century_Sailor/equal_opportunity/Pages/COMMANDCLIMATEASSESSMENT.aspx

Nystrom, David C., *360-Degree Feedback: A Powerful Tool for Leadership Development and Performance Appraisal*, Monterey, Calif.: Naval Postgraduate School, 2001.

Ostroff, Cheri, Leanne E. Atwater, and Barbara J. Feinberg, "Understanding Self–Other Agreement: A Look at Rater and Ratee Characteristics, Context, and Outcomes," *Personnel Psychology*, Vol. 57, No. 2, 2004, pp. 333–375.

Pfau, Bruce N., and Scott A. Cohen, "Aligning Human Capital Practices and Employee Behavior with Shareholder Value," *Consulting Psychology Journal: Practice and Research*, Vol. 55, No. 3, 2003.

Pointer, Beverly A., *Leadership Development: A 360 Degree Approach*, Carlisle Barracks, Pa.: U.S. Army War College, 1996.

Psotka, Joseph, Peter J. Legree, and Dawn M. Gray, *Collaboration and Self Assessment: How to Combine 360 Assessments to Increase Self-Understanding*, Alexandria, Va.: Army Research Institute, ARI-RN-2007-03, 2007.

Reese, Timothy R., *Transforming the Officer Evaluation System: Using a 360-Degree Feedback Model*, Carlisle Barracks, Pa.: U.S. Army War College, 2002.

Riley, Ryan, Josh Hatfield, Jon J. Fallesen, and Katie M. Gunther, *2013 Center for Army Leadership Annual Survey of Army Leadership (CASAL): Army Civilian Leaders*, Fairfax Va.: ICF International Inc., Technical Report 2014-01, 2014.

Ripple, Bryan, "DEOMI Organizational Climate Survey Enhancements Improve Commander's Opportunities to Strengthen Readiness Through Awareness," Defense Equal Opportunity Management Institute, undated. As of January 18, 2015: http://www.deomi.org/PublicAffairs/ClimateSurvey.cfm

Scullen, Steven E., Michael K. Mount, and Timothy A. Judge, "Evidence of the Construct Validity of Developmental Ratings of Managerial Performance," *Journal of Applied Psychology*, Vol. 88, No. 1, 2003.

Secretary of the Navy, *Promotion, Special Selection, Selective Early Retirement, and Selective Early Removal Boards for Commissioned Officers of the Navy and Marine Corps*, Washington, D.C.: Department of the Navy, Secretary of the Navy Instruction 1420.1B, March 28, 2006.

Smither, James W., Manuel London, and Richard R. Reilly, "Does Performance Improve Following Multisource Feedback? A Theoretical Model, Meta-Analysis, and Review of Empirical Findings," *Personnel Psychology*, Vol. 58, 2005, pp. 33–66.

Smither, James W., Manuel London, and Kristin Roukema Richmond, "The Relationship Between Leaders' Personality and Their Reactions to and Use of Multisource Feedback: A Longitudinal Study," *Group and Organization Management*, Vol. 30, No. 2, 2005, pp. 181–210.

Smither, James W., Manuel London, Nicholas L. Vasilopoulos, Richard R. Reilly, Roger E. Millsap, and Nat Salvemini, "An Examination of the Effects of an Upward Feedback Program Over Time," *Personnel Psychology*, Vol. 48, No. 1, 1995, pp. 1–34.

Taylor, Curtis D., *Breaking the Bathsheba Syndrome: Building a Performance Evaluation System That Promotes Mission Command*, Fort Leavenworth, Kan.: U.S. Army Command and General Staff College, 2014. As of December 30, 2014: http://cgsc.contentdm.oclc.org/cdm/ref/collection/p4013coll3/id/3200

Tornow, Walter W., "Perceptions or Reality: Is Multi-Perspective Measurement a Means or an End?" *Human Resource Management*, Vol. 32, No. 2–3, 1993, pp. 221–229.

U.S. Office of Personnel Management, *Support Services Contract Cost Benefit Analysis*, Washington, D.C.: Federal Investigative Services Division, undated. As of February 11, 2015: http://www.hsgac.senate.gov/download/?id=F2CC8B19-A04A-4167-A3BA-443827704BDA

Viswesvaran, Chockalingam, Deniz S. Ones, and Frank L. Schmidt, "Comparative Analysis of the Reliability of Job Performance Ratings," *Journal of Applied Psychology*, Vol. 81, No. 5, 1996.

Walker, Alan G., and James W. Smither, "A Five-Year Study of Upward Feedback: What Managers Do with Their Results Matters," *Personnel Psychology*, Vol. 52, No. 2, 1999, pp. 393–423.

Warr, Peter, and Alan Bourne, "Factors Influencing Two Types of Congruence in Multirater Judgments," *Human Performance*, Vol. 12, No. 3-4, 1999, pp. 183–210.

Williams, James M., *The Surface Warfare Community's 360-Degree Feedback Pilot Program: A Preliminary Analysis and Evaluation Plan*, Monterey, Calif.: Naval Postgraduate School, 2005.

Woehr, David J., M. Kathleen Sheehan, and Winston Bennett Jr., "Assessing Measurement Equivalence Across Rating Sources: A Multitrait-Multirater Approach," *Journal of Applied Psychology*, Vol. 90, No. 3, 2005.